AS in a week

Jim Sharpe and
Joanna Whitehead,
Abbey College, Birmingham
Series Editor: Kevin Byrne

Biology

Where to find the information you need

SUCCESS OR YOUR MONEY BACK

Letts' market leading series AS in a Week gives you everything you need for exam success. We're so confident that they're the best revision books you can buy that if you don't make the grade we will give you your money back!

HERE'S HOW IT WORKS

Register the Letts AS in a Week guide you buy by writing to us within 28 days of purchase with the following information:

- Name
- Address
- Postcode
- Subject of AS in a Week book bought

Please include your till receipt

To make a **claim**, compare your results to the grades below. If any of your grades qualify for a refund, make a claim by writing to us within 28 days of getting your results, enclosing a copy of your original exam slip. If you do not register, you won't be able to make a claim after you receive your results.

CLAIM IF...

You are an AS (Advanced Subsidiary) student and do not get grade E or above.
You are a Scottish Higher level student and do not get a grade C or above.
This offer is not open to Scottish students taking SCE Higher Grade, or Intermediate qualifications.

Registration and claim address:
Letts Success or Your Money Back Offer, Letts Educational, 414 Chiswick High Road, London W4 5TF

TERMS AND CONDITIONS

1. Applies to the Letts AS in a Week series only
2. Registration of purchases must be received by Letts Educational within 28 days of the purchase date
3. Registration must be accompanied by a valid till receipt
4. All money back claims must be received by Letts Educational within 28 days of receiving exam results
5. All claims must be accompanied by a letter stating the claim and a copy of the relevant exam results slip
6. Claims will be invalid if they do not match with the original registered subjects
7. Letts Educational reserves the right to seek confirmation of the level of entry of the claimant
8. Responsibility cannot be accepted for lost, delayed or damaged applications, or applications received outside of the stated registration/claim timescales
9. Proof of posting will not be accepted as proof of delivery
10. Offer only available to AS students studying within the UK
11. SUCCESS OR YOUR MONEY BACK is promoted by Letts Educational, 414 Chiswick High Road, London W4 5TF
12. Registration indicates a complete acceptance of these rules
13. Illegible entries will be disqualified
14. In all matters, the decision of Letts Educational will be final and no correspondence will be entered into

Letts Educational
Chiswick Centre
414 Chiswick High Road
London W4 5TF
Tel: 020 8996 3333
Fax: 020 8743 8390
e-mail: mail@lettsed.co.uk
website: www.letts-education.com

Every effort has been made to trace copyright holders and obtain their permission for the use of copyright material. The authors and publishers will gladly receive information enabling them to rectify any error or omission in subsequent editions.

First published 2000
Reprinted 2001
New edition 2004

Text © Jim Sharpe and Joanna Whitehead 2000
Design and illustration © Letts Educational Ltd 2000

British Library Cataloguing in Publication Data
A CIP record for this book is available from the British Library.

ISBN 1 84315 350 5

Cover design by Purple, London

Prepared by *specialist* publishing services, Milton Keynes
Design and project management by Starfish DEPM, London.

Printed in the UK

Letts Educational Limited is a division of Granada Learning Limited, part of Granada plc.

Cells, Tissues and Organisms

15 mins

Time Yourself

How much do you know?

1 4.25 cm is equal to _____ μm.

2 An electron microscope has a greater magnification and resolution because electrons have a _____ wavelength.

3 Organisms can be classified as viruses, _____ and eukaryotes and the smallest of these are the _____ .

4 The cell surface membrane has a basic structure known as the _____ _____ model. The membrane is _____ wide and is composed of a _____ bilayer with proteins.

5 Substances may enter or leave cells by diffusion, osmosis, facilitated diffusion and _____ _____ . This process requires energy in the form of _____ .

6 Organelles can be isolated by _____ _____ .

7 The nucleus encloses and protects the _____ , which controls the activities of the cell.

8 The inner membrane of the mitochondria is folded into _____ and is the site of the _____ _____ _____ . The matrix is the site of the _____ cycle.

9 The rough endoplasmic reticulum (ER) is covered in _____ and synthesized proteins can be moved through the cell in its cavities.

10 The Golgi apparatus are a stack of flattened _____ , and _____ pinch off from these.

11 In a chloroplast, the light dependant reaction takes place in the _____ and the light independent reaction takes place in the _____ .

12 A _____ is an aggregation of similar cells which perform a particular function.

Answers

1 4.25×10^4 μm **2** smaller **3** prokaryotes, viruses **4** fluid mosaic, 7.5nm, phospholipid **5** active transport, ATP **6** cell fractionation **7** DNA **8** cristae, electron transport chain, Krebs **9** ribosomes **10** cisternae, vesicles **11** thylakoids, stroma **12** tissue

If you got them all right, skip to page 10

Learn the key facts

1 The units of measurement in biology are determined by SI (Systeme Internationale) units. The most commonly used include micrometres and nanometres.

1 metre (m)

1 centimetre	=	1×10^{-2}m (cm)
1 millimetre	=	1×10^{-3}m (mm)
1 micrometre	=	1×10^{-6}m (μm)
1 nanometre	=	1×10^{-9}m (nm)
1 picometre	=	1×10^{-12}m (pm)

2 Biological specimens can be observed using either a light or electron microscope. There are two types of electron microscope, either a scanning or transmission electron microscope. Electrons are reflected from the specimen in an SEM and are passed through a TEM. Magnification is increasing the apparent size of the image and resolution is the ability to distinguish between two points.

	Advantages	Disadvantages
LIGHT	View living specimens and colour	Magnification ($\times 25 - \times 1500$) Resolution (0.2mm apart)
TEM	Magnification ($\times 50000 - \times 500000$) Resolution (1nm apart)	Artefacts appear due to staining techniques
SEM	Magnification (same as TEM) 3D image obtained	Resolution not as good as TEM

3 Organisms can be classified as viruses, prokaryotes and eukaryotes.

	Viruses	Prokaryotes	Eukaryotes
Example	Flu virus, HIV	Bacteria	Plants, animals, fungi, protoctists
Size	10–30nm (nanometres)	0.1–10µm (micrometres)	10–100µm (micrometres)

Nuclear material	Either RNA or DNA plus proteins	Circular DNA, no histone proteins. Not organised within a membrane-bound nucleus	Linear DNA attached to histone proteins. Organised within a membrane-bound nucleus
Internal organisation	No membranes or organelles	No membranes or organelles	Complex membranes compartmentalize the cell

4 The membranes that surround eukaryotic cells and the membranes that form their organelles all have the same basic structure, known as the fluid mosaic model. It is 7.5 nm thick and consists of a lipid bilayer with some proteins embedded in the surface and some running right through it.

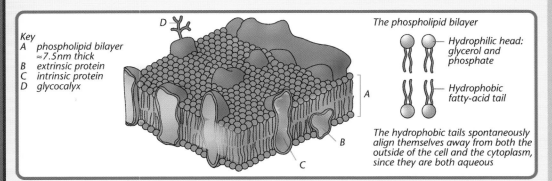

Key
A phospholipid bilayer ≈7.5nm thick
B extrinsic protein
C intrinsic protein
D glycocalyx

The phospholipid bilayer

Hydrophilic head: glycerol and phosphate

Hydrophobic fatty-acid tail

The hydrophobic tails spontaneously align themselves away from both the outside of the cell and the cytoplasm, since they are both aqueous

The interaction between the hydrophobic and hydrophilic ends of the phospholipids gives the membranes stability. Lipids also give the membranes selective permeability. Lipid-soluble (hydrophobic) molecules diffuse through the membrane easily. Hydrophilic substances cross the membrane through channels in the intrinsic proteins. The lipids/proteins can move laterally or change places and this gives the membrane fluidity. This is essential for processes such as endocytosis (movement of material into a cell).

Draw a gap between the phospholipid molecules.

5 Substances may cross the plasma membrane to enter or leave cells by four mechanisms.

Mechanism	Description	ATP required	Example
Diffusion	Movement of molecules from a region of high concentration to low concentration	No	Oxygen, carbon dioxide
Osmosis	Diffusion of water from a high to a low water potential through a differentially permeable membrane	No	Water
Facilitated diffusion	Movement of substances by attachment to transport proteins	No	Glucose
Active transport	Movement of substances against their concentration gradient	Yes	Glucose

6 Within eukaryotic cells, the cytoplasm is divided up by membranes into compartments or organelles. These organelles can be extracted by cell fractionation. The tissue is cut up in a cold isotonic solution – cold to prevent activity of autolytic enzymes and isotonic to prevent lysis (splitting) of organelles. The filtrate is centrifuged by differential centrifugation, initially at low speed and then higher speeds. A sediment forms on the bottom containing the organelle and the liquid above (the supernatant) is spun again at a higher speed. Initially, the cell debris is removed followed in order by nuclei, chloroplasts, mitochondria and finally ribosomes.

7 Each organelle is specialised for a particular function.

Organelle	Structure and function
Nucleus 	A double nuclear envelope encloses and protects DNA (normally visible as chromatin granules).
	Nuclear pores allow entry of substances such as nucleotides for DNA replication and exit of molecules such as mRNA during protein synthesis.
	The outer membrane of the nuclear envelope is continuous with the rough endoplasmic reticulum membranes. This makes the perinuclear space (space within the nuclear envelope) continuous with the lumen of the endoplasmic reticulum, thus allowing easy transport of substances.
8 Mitochondria 	A double membrane isolates reactions of the Krebs cycle and electron transport chain from the general cytoplasm. Such compartmentalization allows high concentrations of enzymes and substrates to be maintained, which increases the rate of respiratory reactions.
	The inner membrane is folded to form cristae, which greatly increase the surface area for the electron transport chain reactions.
	The matrix contains • 70S ribosomes for protein manufacture, DNA for protein manufacture; • enzymes, e.g. decarboxylase used in Krebs cycle.
9 Endoplasmic reticulum 	Endoplasmic reticulum is a system of hollow tubes and sacs which allow transport of substances within the cell.
	Rough endoplasmic reticulum (RER) is covered with ribosomes and consists of an interconnected system of flattened sacs. The ribosomes synthesise proteins which can then be transported through the cell in the cavities of the endoplasmic reticulum. The percentage of RER is high in cells which produce proteins for export, e.g. digestive enzymes.
	Smooth endoplasmic reticulum (SER) – which lacks ribosomes – is a system of interconnected tubules. This is the site of lipid synthesis.
10 Golgi body 	The Golgi body consists of flattened cisternae (membrane-bound cavities) which allows internal transport. Vesicles contain materials to be secreted.
	The Golgi body is connected to the RER. Proteins from the RER are modified before secretion. For example, carbohydrates may be added to proteins to form glycoproteins such as mucus.

11

Organelle	Structure and function
Ribosomes 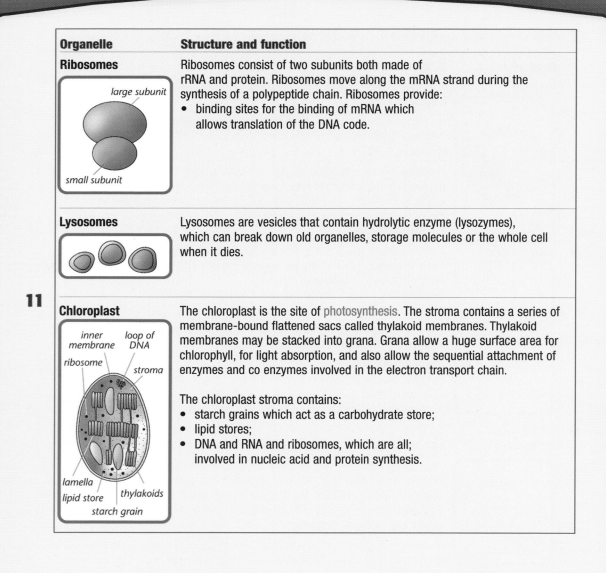	Ribosomes consist of two subunits both made of rRNA and protein. Ribosomes move along the mRNA strand during the synthesis of a polypeptide chain. Ribosomes provide: • binding sites for the binding of mRNA which allows translation of the DNA code.
Lysosomes	Lysosomes are vesicles that contain hydrolytic enzyme (lysozymes), which can break down old organelles, storage molecules or the whole cell when it dies.
Chloroplast	The chloroplast is the site of photosynthesis. The stroma contains a series of membrane-bound flattened sacs called thylakoid membranes. Thylakoid membranes may be stacked into grana. Grana allow a huge surface area for chlorophyll, for light absorption, and also allow the sequential attachment of enzymes and co enzymes involved in the electron transport chain. The chloroplast stroma contains: • starch grains which act as a carbohydrate store; • lipid stores; • DNA and RNA and ribosomes, which are all; involved in nucleic acid and protein synthesis.

12 A tissue is an aggregation of similar cells which perform a particular function or functions. Tissues can be simple tissues (one cell thick) or compound tissues (more than one cell thick).

Simple tissues

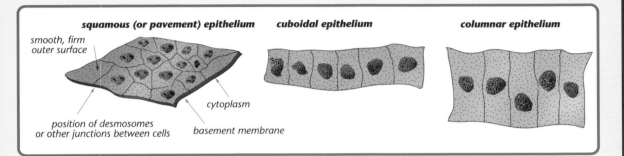

squamous (or pavement) epithelium   cuboidal epithelium   columnar epithelium

smooth, firm outer surface

position of desmosomes or other junctions between cells

cytoplasm

basement membrane

Cells, Tissues and Organisms

Have you improved?

DAY

1

2

3

4

5

6

7

1 Complete the table below which compares cellular transport mechanisms.

	Simple diffusion	Facilitated diffusion	Active transport
Is ATP required?	N		Y
Rate of movement		Fast	
Direction of transport in relation to concentration gradient	Along		Against

2 The diagram shows part of a cell surface membrane.

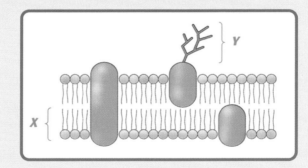

a) Identify structures X and Y.

b) Explain why the cell surface membrane is said to be fluid.

c) Suggest why some liver cells have very high percentages of:
   (i)   SER;
   (ii) mitochondria.

Organic Molecules

How much do you know?

1 Water molecules have two charged ends and can form weak _____ bonds. Water is said to be a _____ molecule.

2 The basic units of carbohydrates are _____, characterised by the number of carbon atoms, e.g. _____ sugars have five carbons.

3 Two monosaccharides can link together to form a _____. The bond between them is a _____ linkage formed in a _____ reaction.

4 Starch is a polysaccharide composed of _____ and amylopectin. Its main function is as a _____ carbohydrate. Starch and glycogen are suitable for this function because they are _____ and compact.

5 A _____ is composed of 1 molecule of glycerol and 3 fatty acids.

6 A triglyceride is formed from a condensation reaction called _____.

7 Fatty acids with double bonds are said to be _____ and have a _____ melting point than fatty acids with no double bonds.

8 Amino acids are _____, acting as both an acid and a base. Through condensation of two amino acids they form _____.

9 The _____ structure of a protein is the sequence of its amino acids. Further structures are controlled by the formation of weak _____ bonds and _____ bridges between amino acids.

10 In chromatography, the amino acids can be identified by calculating the _____ value.

11 If reducing sugars are present in foods, Benedict's solution will turn from blue to a _____ _____ on heating.

Answers

1 hydrogen, dipolar **2** monosaccharides, pentose **3** disaccharide, glycosidic, condensation **4** amylose, storage, insoluble **5** triglyceride **6** esterification **7** unsaturated, lower **8** amphoteric, dipeptides **9** primary, hydrogen, disulphide **10** Rf **11** brick red precipitate

If you got them all right, skip to page 19

Organic Molecules

Learn the key facts

1 Water is a dipolar molecule with two charged ends, which attract ions and other charged molecules forming weak hydrogen bonds. This makes water an excellent solvent as water molecules form shells around the solute molecules, e.g. Na+. It also allows cohesion (water molecules cling together) and adhesion (water molecules stick to other polar molecules).

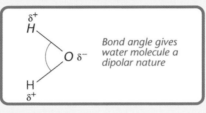

Bond angle gives water molecule a dipolar nature

Water has the following properties:

- Small relative molecular mass (18). It is a liquid at room temperature needed for chemical reactions.
- High specific heat capacity. A large amount of heat energy is needed to raise the temperature of 1 g of water by 1°C . This maintains a constant internal environment where temperature changes are minimised in cells.
- High latent heat of vaporisation. A large amount of heat energy is needed to turn a given amount of liquid water into water vapour. This is used in sweating to cool down the body temperature.
- Ice formation. Ice is less dense than water and floats, which insulates the water below and allows aquatic life to surivive.
- High latent heat of fusion. A large amount of heat needs to be lost before water freezes. Cell contents are less likely to be damaged by ice crystals.
- High surface tension allows movement of small insects.
- Hydrolysis. Water is used to split up large organic molecules into monomers.

2 Carbohydrates all contain carbon, hydrogen and oxygen, with the general formula $C_x (H_2O)_y$. Their three main functions are:

- main respiratory substrate, e.g. glucose
- food storage compounds, e.g. starch
- structural, e.g. cellulose.

> The properties of water are vital to living organisms. It acts as a solvent and a metabolite and prevents rapid temperature change.

Monosaccharides are the basic units (monomers) which make up all other carbohydrates. They are soluble in water, taste sweet and consist of small molecules, characterised by their number of carbon atoms:

- Triose sugars. Three carbon atoms, e.g. phosphoglyceraldehyde formed in the light-independent stage of photosynthesis.
- Pentose sugars. Five carbon atoms, e.g. ribose in RNA and deoxyribose found in DNA.
- Hexose sugars. Six carbon atoms, e.g. fructose found in nectar and fruits to attract insects, and glucose, the main respiratory substrate. Glucose occurs in two forms:

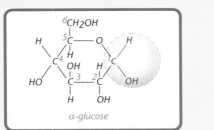

α-glucose

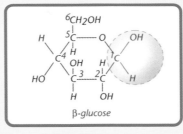

β-glucose

> Notice the difference between a- and b-glucose in the position of the OH and H on carbon atom 1.

3 **Disaccharides** are composed of two monosaccharides joined in a condensation reaction (involving the loss of water) forming a glycosidic linkage, e.g. α-glucose + β-fructose = sucrose.

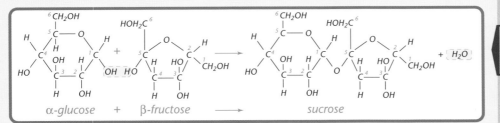

α-glucose   +   β-fructose   ⟶   sucrose

> Remember to show water is removed.

Three common disaccharides are:

- Sucrose (α-glucose + β-fructose). Transports carbohydrate in plant phloem tissue.
- Maltose (α-glucose + α-glucose). Found in germinating seeds.
- Lactose (α-glucose + galactose). Found in milk, providing the energy source for suckling mammals.

DAY

1

4 Polysaccharides are polymers formed by multiple condensation reactions of monosaccharides. They are not sweet or soluble in water and are not crystalline. Their main function in plants and animals is to provide structure and storage because they are insoluble and compact.

Starch	Glycogen	Cellulose
Polymer of α-glucose found in two forms: • Amylose – straight unbranched helix • Amylopectin – branched chain (branching every 24th glucose monomer)	Polymer of branched chains of α-glucose (branching every 10th glucose monomer)	Polymer of β-glucose in an unbranched chain. OH groups stick outwards forming H-bonds with neighbouring chains giving a high tensile strength and permeability
Storage carbohydrate in plants (found in chloroplasts and root cortex parenchyma cells)	Storage carbohydrate in animals, found in muscles and liver, readily hydrolysed to glucose	Form plant cell wall, combined with a gel-like organic matrix

5 Lipids contain carbon, hydrogen and oxygen (although lower amounts of oxygen than carbohydrates) and are compact and insoluble. The general formula is $C_nH_{2n}O_2$. Non-essential fatty acids can be synthesised by the body from carbohydrate and protein metabolism. Essential fatty acids cannot be synthesised and must be obtained from the diet, e.g. vegetables and seed oils.

Lipids have a variety of functions:

* Storage of metabolic fuels in adipose tissue of animals and oils in plants, e.g. castor oil seeds and fruits (lipids contain double the amount of energy of carbohydrates);
* Heat insulation in hibernating and marine mammals, e.g. whales;
* Buoyancy and waterproofing in aquatic animals, e.g. geese.

Distinguish between these polymers.

Storage molecules are insoluble so they do not affect the movement of water across the cell membranes.

6 All lipids are composed of glycerol and fatty acids.

Glycerol Molecular arrangement is the same in all lipids.

Fatty acids The nature of a lipid depends on the particular fatty acids it contains.

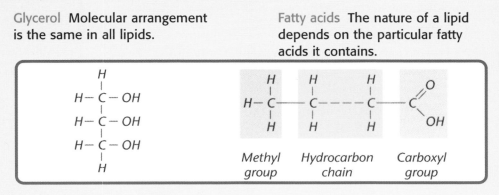

	Methyl group	Hydrocarbon chain	Carboxyl group

The general formula of lipids is $C_nH_{2n}O_2$.

A triglyceride is formed by a condensation reaction called esterification from one molecule of glycerol and three fatty acids. Ester bonds and three molecules of water are formed.

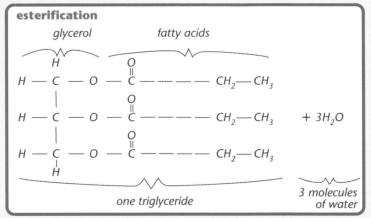

esterification

glycerol   fatty acids

one triglyceride   3 molecules of water

7 A phospholipid has a phosphate group and two fatty acid tails attached to one molecule of glycerol.

Unsaturated fatty acids form oils, e.g. oleic acid in plants, as they have double bonds between carbon atoms which lower the melting point. Saturated fatty acids form fats, e.g. stearic acid in animals, as they have no double bonds, raising the melting points and forming animal fats.

Unsaturated fatty acids do not contain the full number of H atoms.

Lipid formation occurs when three fatty acids combine with one glycerol to form triglycerides by a condensation reaction, with the loss of three water molecules.

8 Proteins contain carbon, hydrogen and oxygen, along with nitrogen and often sulphur and phosphate. They are polymers of 20 types of amino acids.

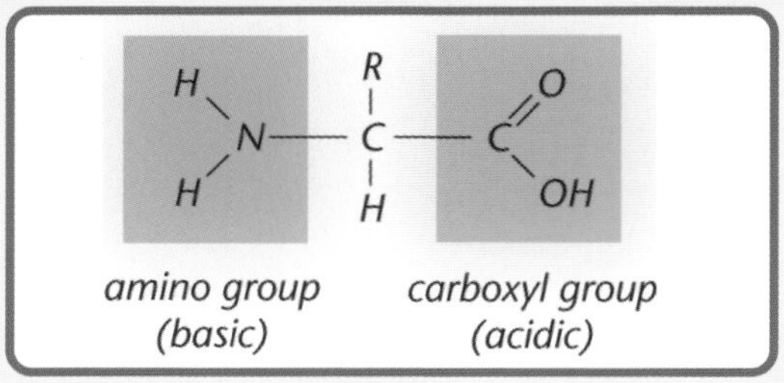

amino group
(basic)

carboxyl group
(acidic)

Amino acids have the following characteristics:

- **Amphoteric.** They act as both an acid and a base depending on surrounding pH. This allows them to act as a buffer to pH change.
- **Zwitterions.** They possess both a negative and positive charge.
- **Form dipeptides** through condensation of two amino acids.

amino acid residue

peptide chain

hydrogen bond

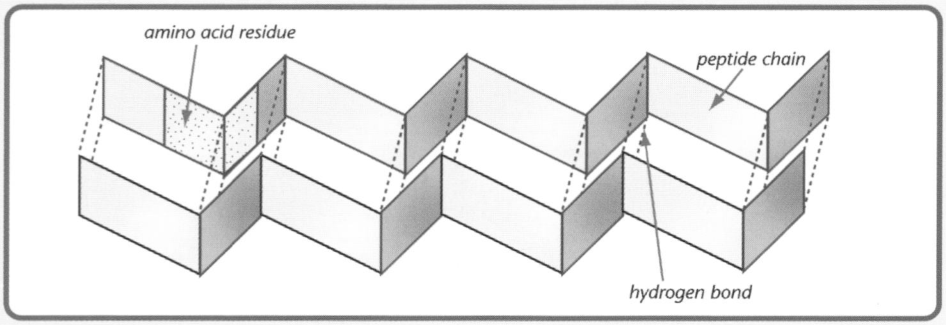

9 Polypeptides are polymers of amino acids joined by peptide bonds.

Protein structure can be divided into four levels:

- **Primary structure.** This is a sequence of amino acids determined by the genetic code of DNA.

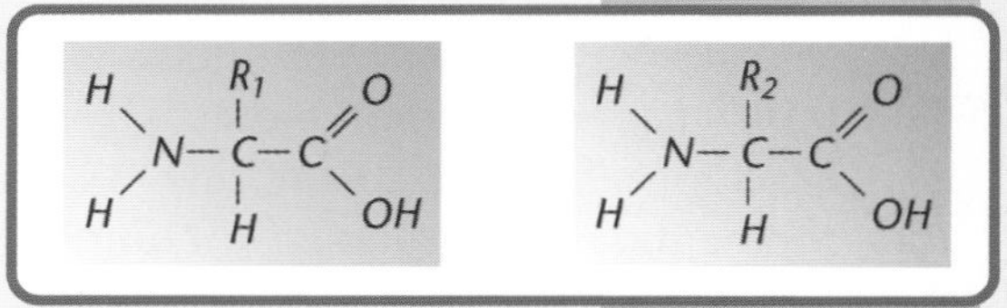

The carboxyl group of each fatty acid bonds to an OH group of glycerol (ester bond).

Define a polymer.

You must know the structure of an amino acid.

DAY

1

2

3

4

5

6

7

- Secondary structure. The polypeptide chain folds either into an α-helix or a β-pleated sheet. These are held together by hydrogen bonds. For example:

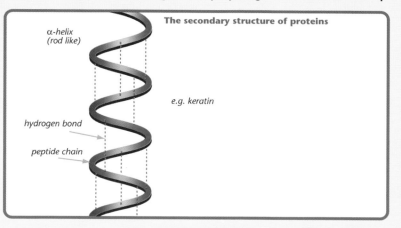

The secondary structure of proteins

α-helix
(rod like)

e.g. keratin

hydrogen bond

peptide chain

- Tertiary structure. The secondary structure (one polypeptide chain) is further folded and is held together by hydrogen and disulphide bonds (formed between two cysteine amino acids). These proteins can either be fibrous proteins which have structural functions and are insoluble, e.g. collagen, or globular proteins which are folded as spherical molecules and are soluble, e.g. enzymes. In globular proteins, the hydrophilic amino acids are found on the outside and the hydrophobic amino acids are found on the inside. This affects the folding of the protein.

- Quarternary structure. These proteins consist of two or more polypeptide chains – e.g. haemoglobin is composed of four polypeptide chains, two α-chains and two β-chains with an attached haem group containing iron.

10 Different amino acids can be separated by paper chromatography. A concentrated spot of the mixture of amino acids is placed on the paper and put into a chromatography tank. The solvent moves up the paper and the solvent front is marked. The paper is dried and sprayed with ninhydrin to see the position of the amino acids, and their Rf values are calculated using the equation:

$$Rf = \frac{\text{distance moved by the amino acid}}{\text{distance moved by the solvent front}}$$

The Rf values identify the amino acids.

11 Food tests identify different organic molecules.

Organic molecule	Reagent	Initial colour	Final colour
Reducing sugars	Benedict's solution	Blue	Brick red ppt
Non-reducing sugars	HCl and Benedict's solution	Blue	Brick red ppt
Starch	Iodine solution	Yellow-brown	Blue-black
Protein	Biuret solution	Pale blue	Purple
Cellulose	Schultz's solution	Yellow-brown	Purple
Lipids	Ethanol and water (emulsion test)	Clear	Milky white suspension

Have you improved?

1 (a) Define the term 'essential fatty acid'.

(b) Give two ways in which phospholipids are different from triglycerides.

(c) State three elements present in fatty acids.

2 For each of the following statements state whether it refers to monosaccharides, fatty acids or amino acids.

(a) Always contain nitrogen.

(b) Produced in the complete hydrolysis of cellulose.

(c) Bonded together by glycosidic links.

(d) Insoluble in water.

3 The diagram below shows two amino acids:

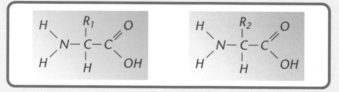

(a) These two amino acids may be linked to form a new molecule.

(i) State the type of reaction that links amino acids and name the molecule that is lost.

(ii) Name the bond formed by this reaction.

(iii) What is the name of the new molecule formed?

(b) Name two elements, other than carbon, hydrogen and oxygen, that may be present in R_1 and R_2.

(c) With reference to the answer you have given in part (b), explain how the presence of different elements at R_1 and R_2 is important in controlling the structure and function of proteins.

Enzymes and Biotechnology

How much do you know?

1 Enzymes are biological catalysts, reducing the _____ energy of reactions. They are composed of _____ proteins with a specific tertiary structure.

2 When a substrate enters an enzyme's active site, the enzyme and substrate form a _____ relationship. In the _____ _____ _____ hypothesis substrate shape is complementary to the shape of the active site.

3 Increasing temperature increases the _____ energy of enzyme and substrate molecules, with high temperature causing the enzyme to become irreversibly _____. Enzymes operate in a _____ pH range, with changing pH around the enzyme disrupting the _____ _____.

4 Increasing enzyme concentration causes a _____ increase in the rate of an enzyme-catalysed reaction, whereas increasing substrate concentration also increases rate, but only until a _____ _____ is reached.

5 Competitive inhibitors bind to the enzyme's _____ site, whereas _____ inhibitors bind elsewhere on the enzyme.

6 Enzymes are used in industry because they allow reactions to occur at normal _____ and _____. For example, clarification of fruit juice occurs through the _____ enzyme.

7 _____ is the process of attaching an enzyme to an inert, solid support. These enzymes are less likely to be denatured by _____ or _____. However, this technique cannot be used for _____ reaction pathways.

Answers

1 activation, globular **2** steric, lock and key **3** kinetic, denatured, narrow active site **4** linear, maximum rate **5** active, non-competitive **6** temperatures, pressures, pectinase. **7** immobilisation, pH, temperature, complex

If you got them all right, skip to page 26

Enzymes and Biotechnology

Spend no more than **45** mins on this topic

Learn the key facts

1 Enzymes control metabolic reactions in the body and the rate at which they occur. Metabolic reactions can be catabolic (breakdown reactions, which release energy overall), or anabolic (building up or synthesising reactions, requiring energy overall).

Enzymes have the following characteristics:

- Biological catalysts, made up of globular proteins with a specific tertiary structure.
- They reduce the activation energy of reactions, which is the energy required for a specific reaction to occur.
- They speed up and slow down metabolic reactions but are never used up in the reaction.

Enzymes, like all catalysts, do not make reactions occur that would not otherwise happen and do not change the volume of product finally produced.

Tertiary structure is the 3D folding of protein chains.

2 The active site is a depression in an enzyme's tertiary structure and is the site of the enzyme-catalysed reaction. The shape of the active site allows only specific substrate molecules to enter, which results in enzyme specificity, with enzymes unique to a specific reaction. The enzyme and substrate are said to form a steric relationship.

Substrate is another name for the reactant molecule.

There are two theories about the action of enzymes:

- Lock and key – Substrate shape is directly complementary to the shape of the active site, with the enzyme acting as the lock and the substrate molecules the key.
- Induced fit – Not all enzymes have a permanent active site. In some, one develops as substrate molecules come close. A small change occurs in the enzyme structure to form a specific active site.

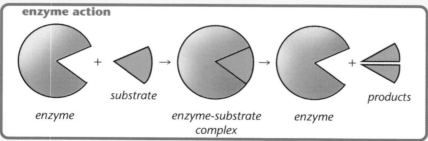

enzyme action

enzyme + substrate → enzyme-substrate complex → enzyme + products

DAY 1 2 3 4 5 6 7

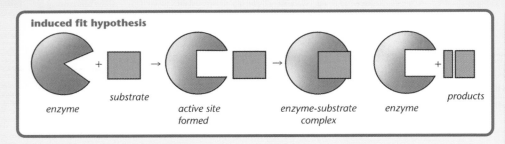

induced fit hypothesis

enzyme + substrate → active site formed | enzyme-substrate complex → enzyme + products

3 The rate of enzyme-catalysed reactions will increase with increasing temperature, up to an optimum rate (the temperature producing the highest reaction rate). Above this optimum, further temperature increase decreases the rate of reaction.

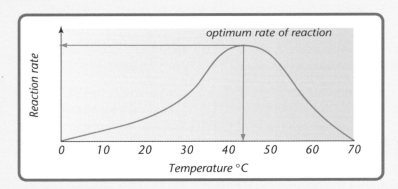

Increasing temperature increases the enzyme and substrate molecule's kinetic energy. The faster-moving molecules collide more often, forming more enzyme–substrate complexes. Temperatures above optimum cause increasing enzyme molecule vibration, breaking down internal bonds and destroying the active site. The enzyme becomes irreversibly denatured.

Enzymes are very sensitive to pH. Individual enzymes operate in a narrow pH range and either side of this optimum irreversible denaturation will occur. Optimum pH for most enzymes is 7 (neutral), although some digestive enzymes, e.g. pepsin, have their optimum rate in acidic conditions.

Changing pH causes changes in the charge of an enzyme's amino acid components. This alters attraction and repulsion forces within the enzyme, disrupting the shape of the active site.

Formation of more enzyme-substrate complexes results in an increased rate of reaction.

Remember, the stomach has a pH of 1.5. Some stomach enzymes are adapted to this.

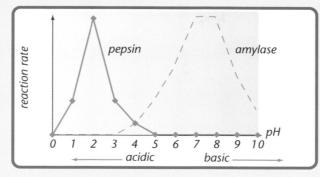

4 The rate of an enzyme-catalysed reaction is also affected by concentration of both enzyme and substrate.

Enzyme concentration	Increasing enzyme concentration causes a linear increase in rate (provided the substrate concentration is not limiting)	Increasing enzyme concentration increases the number of active sites available to catalyse the reaction

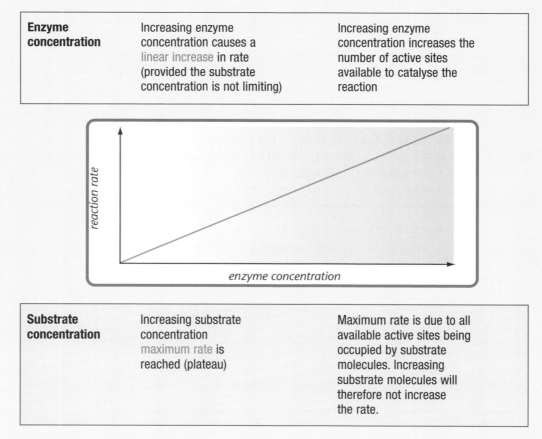

Substrate concentration	Increasing substrate concentration maximum rate is reached (plateau)	Maximum rate is due to all available active sites being occupied by substrate molecules. Increasing substrate molecules will therefore not increase the rate.

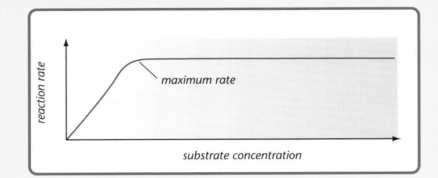

Above max rate, substrate molecules must wait for an active site to become available.

5 Inhibitors are substances that prevent normal enzyme activity. When they are added to a mixture of enzyme and substrate, the rate of reaction is reduced. Inhibitors combine with the enzyme to form an enzyme-inhibitor complex, so the enzyme cannot combine with any substrate molecules.

- Competitive inhibitors compete with the substrate for the active site. The level of inhibition is dependent upon the relative concentrations of substrate and inhibitor, with increasing inhibitor concentration increasing the level of inhibition. This type of inhibition is reversible.

- Non-competitive inhibitors bind to the enzyme, but not at the active site. The level of inhibition depends only on the concentration of the inhibitor, not the substrate, because the inhibitor has a higher affinity for the enzyme than the substrate. This type of inhibition is irreversible.

Heavy metals such as arsenic and mercury are non-competitive inhibitors, binding to sulphur groups present.

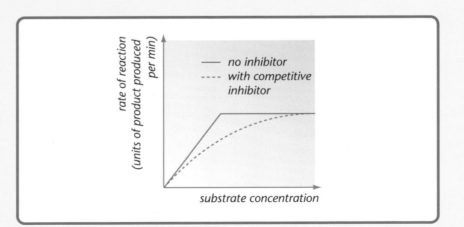

6 Enzymes are widely used in industry because they can accomplish reactions at normal temperatures and pressures, saving the cost of expensive, energy-demanding high temperatures and pressures. The enzymes may be extracted from cells and purified or used within cells.

Enzyme	Industry	Application
Pectinase	Fruit juice manufacture	Clarification of fruit juice
Proteases	Biological detergent manufacture	Used for pre-soaking and direct stain removal

Problems may arise in using enzymes for detergents as people develop allergic reactions. This can be overcome by preventing the enzyme coming into direct contact with people (encapsulation techniques).

7 Immobilisation is the process of attaching or trapping an enzyme to an inert, solid support, e.g. agar gels or cellulose.

Advantages	Disdvantages
Enzymes can be reused, saving the cost of continual production. Enzymes do not contaminate the product. Enzymes are less likely to be denatured by pH or temperature.	Initial cost of immobilisation is high. Technique cannot be used for complex reaction pathways.

Immobilised lactase is used to produce lactose-reduced milk in the dairy industry. Lactase is bound to alginate beads and milk is poured over the beads, producing milk with lactose broken down to glucose and galactose.

Enzymes and Biotechnology

Have you improved?

1 (a) Explain how the shape of an enzyme molecule enables specificity.

(b) Name three factors that affect enzyme activity.

(c) Explain the following terms in relation to enzyme activity:

(i) active site;

(ii) competitive inhibition.

> How do enzyme reactions occur?

2 An experiment was conducted to investigate the activity of the enzyme catalase in potatoes. Ground-up potato tissue (with buffer solution) was mixed with a substrate in a conical flask and the volume of oxygen bubbles produced was measured.

(a) Explain why a buffer solution is mixed with the catalase.

(b) State two reasons for grinding up the potato tissue.

(c) Suggest a name for the substrate on which catalase acts.

(d) State and explain two other pieces of apparatus that should be used when conducting this experiment.

> What factors will affect enzyme activity?

3 (a) The use of enzymes is becoming increasingly widespread in many industrial processes.

(i) Name and state the role of an industrial enzyme.

(ii) Suggest why the use of enzymes is becoming increasingly widespread in industrial processes.

(b) Using immobilised enzymes is one technique available when using enzymes in industry.

(i) Define enzyme immobilisation.

(ii) Describe two advantages and two disadvantages of using immobilised lactase to produce lactose free milk.

Respiration and Gaseous Exchange

15 mins

Time Yourself

How much do you know?

1 Respiratory organs have specific characteristics which increase their effectiveness. These include being _____ so gases can dissolve, having a _____ surface to minimise diffusion distance and having a large surface area.

2 Gaseous exchange takes place in the _____ _____ _____ and the _____ in leaves.

3 _____ is the movement of air in and out of the lungs. This occurs by changing the volume of the _____. During _____, the external intercostal muscles contract, moving the rib cage up and out, and the diaphragm contracts.

4 Exhaled air contains _____% oxygen, which is used to resuscitate an unconscious person.

5 Increased cellular respiration _____ blood CO_2 concentration and _____ blood pH. This is detected by chemoreceptors. The _____ _____ sends nervous impulses to intercostal muscles and the diaphragm to speed up the rate of inspiration and expiration.

6 _____ are a series of hollow tubules providing the respiratory system of insects. Contraction of thoracic and abdominal muscles flattens and _____ the volume of the body. This _____ pressure forces air _____ _____ the tracheae system. Relaxation of the muscles brings air in.

7 In fish, _____ occurs when the mouth closes and floor of the mouth cavity raises. This reduces volume and increases pressure, forcing water out through the _____. This is known as the _____ pressure pump.

8 Exercise increases the _____ ventilation, which is a product of the tidal volume and _____ rate.

Answers

1 moist, thin **2** spongy mesophyll layer, stomata **3** ventilation, thorax, inspiration **4** 16 **5** increases, decreases, medulla oblongata **6** tracheae, decreases, increased, out of **7** expiration, operculum, buccal **8** pulmonary, breathing

If you got them all right, skip to page 32

DAY

2

Respiration and Gaseous Exchange

Learn the key facts

1 **Respiration** is the intake of oxygen and output of carbon dioxide between the body and air. Respiratory organs have the following characteristics which increase their effectiveness:

- **Internalised gaseous exchange surfaces** minimise water loss.
- **Large surface area** to maximise exchange of gases. Lungs contain over 750 million alveoli (tiny sacs) giving a surface area of 80m^2.
- **Moist surface** because gases must dissolve before diffusion can occur.
- **Thin surface** to maximise diffusion rate by minimising distance to travel. Alveoli are only one cell thick. Diffusion occurs by Fick's Law:

  $$\text{Diffusion rate} = \frac{\text{surface area} \times \text{difference in concentration}}{\text{thickness of exchange surface}}$$

- **Transport mechanism** to carry gases between the respiratory organs and body cells, e.g. mammals use the respiratory pigment haemoglobin in the blood, which has a high affinity for oxygen.
- **Ventilation** to maintain air flow to respiratory organs.

The lungs are the respiratory organs in humans.

> Remember, O2 is needed for cellular respiration – production of ATP.

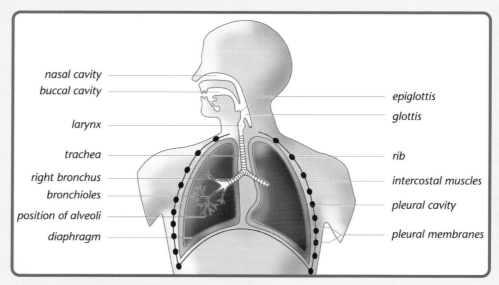

nasal cavity
buccal cavity
larynx
trachea
right bronchus
bronchioles
position of alveoli
diaphragm

epiglottis
glottis
rib
intercostal muscles
pleural cavity
pleural membranes

Respiration and Gaseous Exchange

2 The main gaseous exchange surfaces in dicotyledonous plants are the spongy mesophyll layer and stomata in leaves. Gaseous exchange also occurs in the root hair cells and lenticels in bark.

3 Air is moved in and out of the lungs (ventilation) by changing the volume of the thorax (body cavity containing the lungs), which changes the air pressure in the lungs.

Inspiration (air moving in)	Expiration (air moving out)
1 External intercostal muscles contract, internal intercostal muscles relax, moving rib cage up and out 2 Diaphragm contracts (flattens down)	1 External intercostal muscles relax, internal intercostal muscles contract and ribcage goes down and in under its weight 2 Diaphragm relaxes (domes upwards)
Increased volume of the thorax decreases pressure below atmospheric pressure, so sucking air in	Decreased volume of the thorax increases pressure above atmospheric pressure, so forcing air out

4 The percentage composition of inhaled and exhaled air is different.

	Inhaled	Exhaled
Nitrogen	78%	78%
Oxygen	21%	16%
Carbon dioxide	0.04%	4%

5 Ventilation rate varies depending on level of activity:

- Exercise causes a demand for more ATP, meaning there is an increase in cellular respiration.

- Increased cellular respiration increases blood CO_2 concentration, which decreases pH.

- Chemoreceptors in the pulmonary artery, carotid arteries and medulla oblongata of the brain detect this change in pH.

- **Medulla oblongata** sends nervous impulses to intercostal muscles (via the intercostal nerve) and the diaphragm (via the phrenic nerve) to speed up the rate of inspiration and expiration and the volume of air breathed.

- **Stretch receptors** in the walls of air passages are active during inspiration. Impulses from these inhibit the medulla and prevent over-expansion of the lungs.

CO_2 dissolves in the blood plasma, producing carbonic acid.

6

The respiratory system of insects consists of tracheae (hollow tubules) linking the spiracles (openings along the insect's body) to body cells. Tracheae branch to smaller tracheoles, allowing gaseous exchange throughout the body.

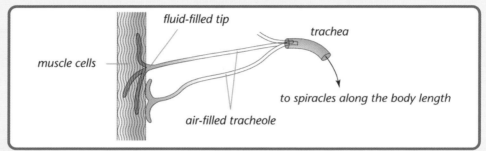

fluid-filled tip

trachea

muscle cells

to spiracles along the body length

air-filled tracheole

7

In fish, gills provide the respiratory organs.

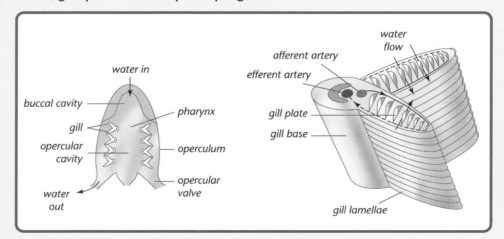

water in

water flow

afferent artery

efferent artery

buccal cavity

pharynx

gill

opercular cavity

operculum

gill plate

gill base

opercular valve

water out

gill lamellae

Water flows over the gills, allowing gaseous exchange when water pressure in the opercular cavity is lower than in the mouth cavity.

Inspiration (opercular suction pump)	Expiration (buccal pressure pump)
• Mouth opens and floor of buccal cavity lowers • Operculum bulges outwards	• Mouth closes and floor of the mouth cavity raises • Muscular contractions force the operculum inwards
Increased volume reduces pressure, sucking water in through the mouth and then over gills	Reduced volume increases pressure, forcing water out through operculum flowing over gill surface

As much O_2 as possible is transferred from the water to the blood in the gill lamellae because the blood flows in the opposite direction to the water flow. This is known as the countercurrent flow.

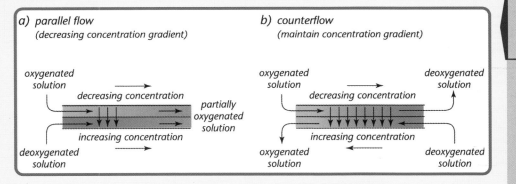

a) parallel flow (decreasing concentration gradient)

b) counterflow (maintain concentration gradient)

The diagram shows that with counterflow, as oxygenated water loses O_2 (reduces O_2 concentration) it encounters blood with a decreasing concentration of O_2, so maintaining a concentration gradient.

8 Exercise increases the pulmonary ventilation, which is a product of the tidal volume and breathing rate. The high concentration of carbon dioxide is detected in the cardiovascular centre in the brain, and nerve impulses pass along the sympathetic nerve to increase heart rate. The amount of blood flowing through the heart is called the cardiac output and this is equal to stroke volume × heart rate.

DAY

2

Counterflow allows high O2 saturation of blood.

30 mins

Time Yourself

Respiration and Gaseous Exchange

Have you improved?

DAY 2

1 (a) Explain why mammals have a complex system for gaseous exchange, whereas very small organisms have no special adaptations for gaseous exchange.

(b) State three characteristics of a gaseous exchange surface.

Remember surface area to volume ratio.

2 (a) Describe how a molecule of carbon dioxide outside the leaves of a flowering plant reaches the spongy mesophyll cells inside a leaf and diffuses into the cytoplasm of the cell.

(b) State and describe two sites of gaseous exchange in plants, apart from stomata.

3 (a) Describe the route taken by a molecule of oxygen from outside the mouth to reaching the alveoli.

(b) The bronchioles contain rings of cartilage. What is the importance of this tissue in the functioning of the respiratory tract?

What happens to volume with pressure change?

4 The diagram below is a representation of an air sac in the lungs showing alveoli:

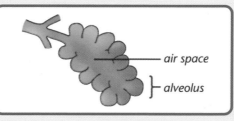

air space

alveolus

(a) How is the structure of the pavement epithelium cells lining the air sac adapted to their function?

(b) Explain why the surface of the alveolus is kept moist.

Protein Synthesis and Nuclear Division

15 mins

Time Yourself

How much do you know?

1 Transcription involves the production of a _____ of genetic code and _____ is the conversion of the template into a polypeptide.

2 Transcription produces _____ controlled by the enzyme DNA-dependent RNA polymerase, which _____ the DNA α-helix.

3 Translation requires two nucleic acid molecules: _____ molecules to carry amino acids, and _____ to translate the mRNA strand.

4 Within the ribosome, _____ tRNA molecules bind to the mRNA allowing amino acids to form a _____ bond.

5 Mitosis is involved in _____ reproduction, since the two resultant daughter nuclei produced are genetically _____ to the parent nuclei.

6 During _____ the chromosomes become attached to spindle fibres.

7 Meiosis produces _____ nuclei from a diploid parental nucleus. The cells produced form _____ responsible for sexual reproduction.

8 In prophase I, homologous chromosomes pair up to form _____.

9 Anaphase II is the separation of the daughter _____ by pulling apart centromeres, producing _____ haploid cells.

10 Meiosis increases genetic variation through _____ and the _____ arrangement of chromosomes during metaphase I.

11 PCR can be used in _____ science and _____ diagnosis.

12 The bar code pattern produced in genetic fingerprinting is called a _____ blot.

Answers

1 template, translation **2** mRNA, unzips **3** tRNA, rRNA **4** two, peptide **5** asexual, identical **6** metaphase **7** haploid, gametes **8** bivalents **9** chromatids, four **10** chiasmata, independent **11** forensic, disease **12** Southern

If you got them all right, skip to page 40

DAY

3

1 2 4 5 6 7

Protein Synthesis and Nuclear Division

Learn the key facts

1 Protein synthesis is the production of polypeptide chains based on the genetic code in DNA. It occurs in two stages in all cells containing nuclei. Transcription is the production of a template of the relevant genetic code and translation is the use of this template to produce a polypeptide chain.

Transcription occurs in the nucleus. It is the production of a chain of the nucleic acid mRNA from DNA under the control of the enzyme DNA-dependent RNA polymerase with ligase activity.

> DNA and RNA are both nucleic acids composed of nucleotide units.

2 RNA polymerase recognises and binds at the start of the relevant DNA section (coding strand). The polymerase now travels along the coding strand, breaking the H-bonds and unwinding the DNA α-helix. This exposes bases of the DNA nucleotides, and RNA nucleotides with complementary bases come in to form the single mRNA strand.

> mRNA is complementary to the DNA strand.

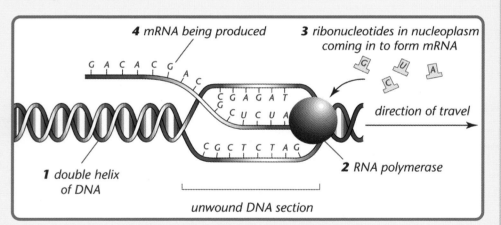

4 mRNA being produced

3 ribonucleotides in nucleoplasm coming in to form mRNA

G A C A C G
A C
C G A G A T
G C U C U A
direction of travel

C G C T C T A G

1 double helix of DNA

2 RNA polymerase

unwound DNA section

The polymerase detaches when it reaches and recognises the stop codon on the DNA. The DNA rewinds and hydrogen bonds reform. The strand of mRNA now leaves the nucleus via a pore into the cytoplasm.

> You must know the base pairing rule.

3 Translation occurs in the cytoplasm and requires tRNA molecules (to bring in amino acids) and a ribosome (to 'translate' the sequence of bases on the mRNA into a polypeptide chain):

DAY
1
2
3
4
5
6
7

- Ribosomes are composed of 50% protein and 50% rRNA, and are made of two subunits which remain separate until mRNA is present.

- tRNA is a single strand of RNA, coiled into a clover-leaf shape.

The two ribosome subunits lock onto the ribosome binding site on the mRNA strand. Each ribosome covers two tRNA binding sites on the mRNA.

tRNA molecules bind to the mRNA strand via complementary bases – called anticodons – to those on the mRNA – called codons. The anticodon corresponds to a specific amino acid brought in by the tRNA.

4 Two tRNA molecules bind to the mRNA within the ribosome and the amino acids carried on the tRNA form a peptide bond together. The first tRNA then disconnects from the mRNA leaving a binding site empty. The ribosome moves along one codon and a third tRNA molecule binds carrying another amino acid to bind to the developing chain. The ribosome continues moving until a stop codon is reached.

> Each mRNA code brings in a rTNA with a specific amino acid.

5 Mitosis is the division of a nucleus (nuclear division) following chromosome replication. The two resultant daughter nuclei have the same number of chromosomes as the parent nuclei (genetically identical). Mitosis is involved in asexual reproduction, and in growth and repair of body tissues.

6 • The first stage is prophase. Chromosomes slowly condense and become visible, and the microtubules break down making the cell spherical. Chromosomes have already replicated and are visible as daughter chromatids joined at a centromere. Microtubules, called spindle fibres, develop from the centrioles (only present in animal cells).

> Plant cells do not contain centrioles, but spindle fibres do form.

- Prometaphase is the start of metaphase. The nuclear envelope breaks down and the chromosomes become attached to spindle fibres, at regions called kinetochores, lining up along the equator of the cell. By the end of metaphase, all the chromosomes have lined up along the metaphase plate, with the spindles pulling towards the poles.

- Anaphase begins when the kinetochores split and the daughter chromatids are pulled to opposite poles of the cell by spindle fibres.

- Telophase can be considered as the reverse of prophase. Chromosomes begin to recondense and new nuclear membranes reform around the two sets of chromosomes.

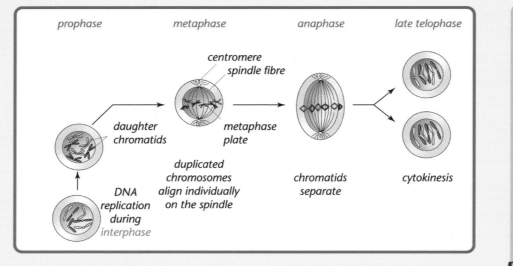

prophase   metaphase   anaphase   late telophase

centromere
spindle fibre

daughter
chromatids

metaphase
plate

duplicated
chromosomes
align individually
on the spindle

chromatids
separate

cytokinesis

DNA
replication
during
interphase

Cytokinesis is cytoplasm division. It is not part of nuclear division but begins when this is complete. Filaments of the proteins actin and myosin contract to produce a cleavage furrow at the equator of the cell and continue until two new cells are produced.

7 Meiosis is also nuclear division following replication of chromosomes, but involves two successive divisions. The four resultant daughter nuclei are haploid (half as many chromosomes as the diploid parental cells).

Meiosis occurs in gamete production in organisms that reproduce sexually. In humans, normal cells contain 46 chromosomes (23 homologous pairs). Gametes contain 23 unpaired chromosomes, so that the fusion of gametes at fertilisation restores the diploid number.

8 • In prophase I, chromosomes condense and homologous pairs come together, forming bivalents. The bivalents are linked at points along their length (chiasmata) and this linkage results in crossing over of genetic material between chromosomes. Chiasmata are random and result in genetic recombinants (new combinations of genes).

• Metaphase I is identical to metaphase in mitosis, except the chromosomes line up as bivalents on the spindle, via the kinetochore fibres.

> Diploid cells contain a complete set of chromosomes.

DAY
1
2
3
4
5
6
7

- The homologous chromosomes separate during anaphase I, with one member of each pair moving to the poles by sliding of kinetochores.

- Telophase I is the final stage of the first division and is the reverse of prophase. Cytokinesis occurs, producing two cells containing one member of each homologous pair of chromosomes. Each chromosome is still made up of two daughter chromatids.

Division 1 separates the homologous pairs of chromosomes.

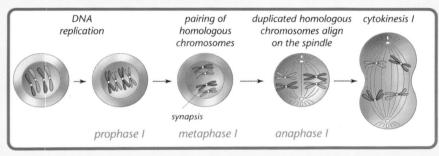

DNA replication · pairing of homologous chromosomes · duplicated homologous chromosomes align on the spindle · cytokinesis I

synapsis

prophase I · metaphase I · anaphase I

9 The second meiotic division occurs in both cells produced during division I. In prophase II chromosomes attach to spindle fibres. The chromosomes then line up individually on the spindle fibre at the equator of the cell in metaphase II. Anaphase II is the separation of the daughter chromatids by pulling apart centromeres. Telophase II produces four haploid cells that are genetically different, following cytokenesis.

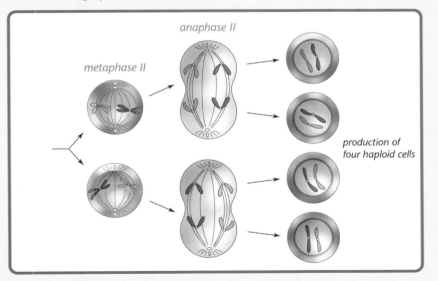

anaphase II

metaphase II

production of four haploid cells

10 Meiosis is significant because it conserves the number of chromosomes during sexual reproduction. If a gamete of an organism had the diploid number of chromosomes, the number of chromosomes would double every generation.

Meiosis also increases genetic variation in two ways:

- The number of possible chromosome combinations is very large in a gamete, due to the independence of chromosome arrangement on the spindle at metaphase I.

- Chiasmata causes recombination of paternal and maternal genetic material in the gametes, increasing variation in the population.

> Fusion of two haploid gametes gives one diploid zygote.

Cells spend only 10% of their lifetime in nuclear division, and 90% in interphase. During interphase the cell is metabolically active, synthesising new proteins and DNA. The chromosomes are uncondensed forming fine threads. The threads consist of nucleosomes, short lengths of DNA wound around a 'bead' of histone proteins.

In most organisms, differentiated tissue cells lose the ability to divide. When they die, these cells are replaced by division from undifferentiated stem cells.

11 The polymerase chain reaction is used to make identical copies of lengths of DNA. It is a series of cyclic reactions performed under sterile conditions. The initial sample of DNA is suspended in a buffer with:

1. Deoxynucleotide triphosphate (dNTPs) – nitrogenous bases to synthesis new DNA

2. Taq DNA polymerase – heat stable DNA polymerase

3. DNA primers – complementary to the DNA either side of the area to be copied.

The PCR reactions occur in a thermal cycler, which can heat up and cool down rapidly to produce an accurate temperature cycle. Approximately 30 cycles are required to produce 1 million copies of a single strand of a DNA molecule.

PCR has many uses, for example:

- Forensic science – PCR amplifies small quantities of DNA found at crime scenes, e.g. blood or semen, allowing genetic fingerprinting to be carried out.

- Disease diagnosis – PCR amplifies genetic material from bacteria and viruses in blood to diagnose disease at an early stage, when only a small number of pathogens are present.

12 Genetic fingerprinting is a technique used in forensic science to reveal unique patterns in an individual's DNA. Particular sections of DNA are repeated thousands of times at different positions along the strand. A radioactive probe of such sequences produces the 'bar code pattern' of different bands – a Southern blot. These repeated sequences have lengths and patterns unique to individuals. Applications of genetic fingerprinting include criminal investigation and researching family trees.

1 The diagram below shows translation occurring on a ribosome:

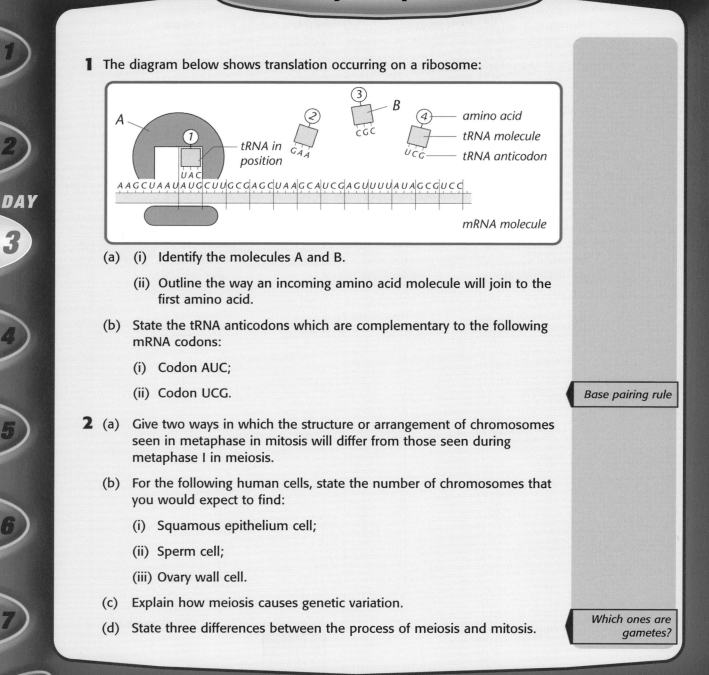

(a) (i) Identify the molecules A and B.

   (ii) Outline the way an incoming amino acid molecule will join to the first amino acid.

(b) State the tRNA anticodons which are complementary to the following mRNA codons:

   (i) Codon AUC;

   (ii) Codon UCG.

Base pairing rule

2 (a) Give two ways in which the structure or arrangement of chromosomes seen in metaphase in mitosis will differ from those seen during metaphase I in meiosis.

(b) For the following human cells, state the number of chromosomes that you would expect to find:

   (i) Squamous epithelium cell;

   (ii) Sperm cell;

   (iii) Ovary wall cell.

(c) Explain how meiosis causes genetic variation.

(d) State three differences between the process of meiosis and mitosis.

Which ones are gametes?

Nutrition

15 mins
Time Yourself

How much do you know?

1 _____ organisms feed on complex _____ foods to obtain the nutrients they require.

2 Digestion occurs through _____ and chemical breakdown. Food is moved along the gut by _____.

3 Pepsin is produced from gastric glands in the _____. It is an _____ which breaks down the _____ peptide bonds.

4 Products of digestion are absorbed across the ileum by _____ _____ and active transport. The ileum is highly adapted for absorption with finger-like projections, called _____, and a good blood supply to maintain the _____ gradient to take up food.

5 Vitamin ___ is fat soluble and is needed to absorb calcium. The deficiency disease is called _____.

6 Haemoglobin contains _____, which is found mainly in _____ _____. The deficiency disease is called _____.

7 *Rhizopus* fungi break down _____ organic material by _____ digestion and absorb the digested food. This is known as _____ nutrition.

DAY
3

Answers

1 heterotrophic, organic **2** mechanical, peristalsis **3** stomach, endopeptidase, internal **4** facilitated diffusion, villi, concentration **5** D, rickets **6** iron, red meat, anaemia **7** dead/decaying, external, saprophytic

If you got them all right, skip to page 46

Spend no more than **45 mins** on this topic

Learn the key facts

1 **Heterotrophic nutrition** involves obtaining molecules needed for cellular metabolism by breaking down complex organic molecules, obtained by feeding on other living organisms. New organic molecules can then be synthesised. In mammals, digestion occurs in the gut or alimentary canal.

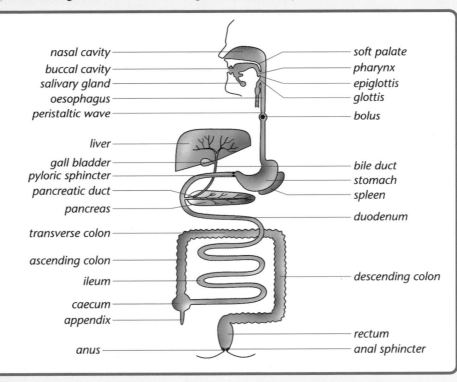

nasal cavity
buccal cavity
salivary gland
oesophagus
peristaltic wave
liver
gall bladder
pyloric sphincter
pancreatic duct
pancreas
transverse colon
ascending colon
ileum
caecum
appendix
anus

soft palate
pharynx
epiglottis
glottis
bolus
bile duct
stomach
spleen
duodenum
descending colon
rectum
anal sphincter

You must be able to label the mammalian gut.

2 Digestion occurs through **mechanical breakdown** (chewing food and contractions of the gut wall), which increases the surface area for **chemical breakdown** by enzymes.

The food is moved along the entire gut by **peristalsis**, which occurs by the action of **antagonistic** muscles. These have opposing effects, i.e. when the circular muscle is contracting, the longitudinal muscle is relaxing.

Food is digested into its soluble constituents to allow absorption into the blood.

DAY 3

3 The following table summarises the action of enzymes in the gut.

Digestive action			
Enzyme	**Site of production**	**Substrate**	**Product(s)**
Salivary amylase	Parotid salivary gland (mouth)	Starch	Maltose
Endopeptidases e.g. pepsin	Gastric pits in stomach (pepsin secreted as pepsinogen, activated by HCL)	Proteins	Polypeptide
Exopeptidases e.g. trypsin	Pancreas (acting in duodenum)	Polypeptides	Amino acids
Pancreatic amylase	Exocrine region of pancreas (acting in duodenum)	Starch	Maltose
Pancreatic lipase	Pancreas	Lipids	Fatty acids and glycerol
Lactase	Cells in crypts of Lieberkühn of duodenum (small intestine)	Lactose	Glucose and galactose
Maltase		Maltose	Glucose
Sucrase		Sucrose	Glucose and fructose

An endopeptidase is a protein-digesting enzyme which breaks down internal peptide bonds, whereas an exopeptidase breaks down terminal peptide bonds.

4 When digestion is complete the products are absorbed across the ileum via facilitated diffusion and active transport.

Active transport allows food to be absorbed against a concentration gradient.

DAY

3

The ileum is highly adapted for absorption:

• Single layer of cells of the epithelium provides a short distance for uptake of digested food.
• Folds in the wall called microvilli increase surface area. These folds have finger-like projections (villi), the cells of which have microvilli.
• Good blood supply maintains the concentration gradient, rapidly removing uptaken food.

• In adult humans the small intestine is 6m long, which provides a large surface area.

5 Vitamins are essential in minute quantities for humans to remain healthy.

Vitamin	Solubility	Function	Deficiency	Food source
A	Fat	Vision	Night-blindness	Green vegetables
B_1	Water	Release of energy from carbohydrates	Beri-beri	Bread
C	Water	Connective tissue	Scurvy	Citrus fruits
D	Fat	Absorbs calcium	Rickets	Butter
K	Fat	Clotting of blood	Blood fails to clot	Dark green leafy vegetables

6 Inorganic ions called minerals are also needed in minute quantities for humans to remain healthy.

Mineral	Function	Deficiency	Food source
Calcium	Hard bones and teeth	Stunted growth	Milk, cheese
Iron	Haemoglobin	Anaemia	Meat
Iodine	Needed for hormones produced in thyroid	Hypothyroidism	Sea-foods
Phosphorus	ATP, phospholipids DNA, RNA, NAD	Stunted growth	All foods
Sodium	Nerve impulses	Muscular cramps	Salt

7 Saprophytic nutrition is a type of heterotrophic nutrition whereby organisms, e.g. *Rhizopus* fungi, obtain their organic molecules from dead or decaying plant and animal material, by releasing enzymes externally (extracellular digestion) and absorbing the digested food.

Parasitic nutrition involves obtaining organic molecules by living on (ectoparasites) or in (endoparasites) a host organism, causing some degree of harm. For example, *Taenia*'s (tape worm) primary host is humans, where it lives in the intestine absorbing the pre-digested food.

DAY

3

Have you improved?

1 Shruti prepared a starch-agar plate to grow three different strains of fungus. The plate was flooded with iodine solution and the results after a week are shown below.

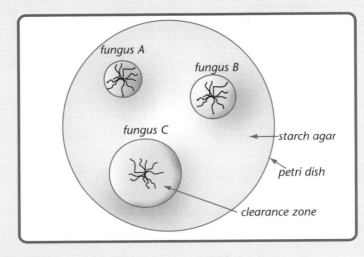

   (a) Why are clearance zones present around each fungus?

   (b) Which fungal strain produced the largest concentration of enzyme?

   (c) Explain how you could set up a control.

2 (a) Identify the region of human digestive system where each of the following occurs:

     (i)   production of hydrochloric acid;

     (ii) emulsification of fats;

     (iii) facilitated diffusion of glucose molecules.

   (b) Describe three ways in which the structure of the ileum is adapted to the function it performs.

   (c) Explain the role of the smooth muscle in the wall of the ileum.

Transport in Mammals

How much do you know?

1 In mammals, _____ carry blood under high pressure and veins carry blood under low pressure.

2 Erythrocytes are adapted to carry _____ by their _____ _____ shape and red pigment _____.

3 Haemoglobin takes up O_2 least readily when fully _____, and fully oxygenated haemoglobin dissociates _____ readily.

4 The _____ curve for haemoglobin shows it can become fully saturated with O_2 at relatively low partial pressures, i.e. in the _____.

5 If blood pH decreases, the oxygen-dissociation curve shifts to the _____, because the haemoglobin has a _____ affinity for O_2.

6 Mammals have a _____ circulatory system, as blood passes through the heart twice before being pumped around the body. The two parts of the circulatory system are the _____ and systemic.

7 There are two groups of _____, or white blood cells: _____ with a granular cytoplasm and agranulocytes.

8 Following atrial systole, the _____ node is stimulated and the impulse is spread to the apex of the ventricles via conducting tissues in the _____ _____ _____.

9 The cardiac-control centre of the _____ _____ in the hind brain receives information on blood pressure. The sympathetic ANS secretes _____ which speeds up the heart rate. The _____ system secretes acetylcholine which slows down heart rate, acting on the SAN or vagus nerve.

10 Tissue fluid forms due to high _____ pressure and then water returns to the capillary under osmotic pressure exerted by _____ proteins.

Answers

1 arteries **2** oxygen, biconcave disc, haemoglobin **3** deoxygenated, most **4** dissociation/sigmoid, lungs **5** right, reduced **6** double, pulmonary **7** leucocytes, granulocytes **8** atrioventricular, Bundle of His **9** medulla oblongata, noradrenaline, parasympathetic **10** hydrostatic, plasma

If you got them all right, skip to page 53

Learn the key facts

1 Large multi-cellular organisms need a transport system to supply every cell with oxygen, water and nutrients. This is because large organisms have a small surface area to volume ratio, so the body surface cannot be used for gaseous exchange because diffusion is too slow to meet demand. Animals use a system of specialised tubes to supply their cells.

vessels	characteristics	function	
Arteries Carry blood away from the heart	• Lined with smooth epithelial cells • Thick muscular, elastic walls with the the ability to stretch and recoil • Lumen 100 µm → 3 cm	• Offer low resistance • Can withstand high pressure • Able to maintain continuous pressure	*5mm*
Capillaries Arranged in beds forming site of exchange to cells	• Lumen 8 µm with wall thickness 2 µm • Pores in capillary wall (with basement membrane) • Numerous, forming large cross-sectional area	• Short diffusion distance between cells and blood • Pores allow production of lymph (bathing cells)	*0.005mm*
Veins Carry blood back to heart	• Less elasticity and muscle fibre than arteries • Pocket-like folds (valves), strengthened by fibrous tissue • Lumen 30 µm → 2.5cm	• Blood flow is assisted by movement of skeletal muscles • Valves prevent backflow of low pressure blood	*5mm*

Arteries are adapted to resist high pressure, veins to prevent back flow.

2 Mammalian blood is 55% plasma and 45% cells. Plasma consists of water, proteins and inorganic ions, e.g. Na^+. Its functions are:

- Maintaining blood pressure.

- Transport of (i) products of digestion, e.g. amino acids and glucose, (ii) metabolic waste products, e.g. CO_2 and urea, (iii) hormones.

Blood cells comprise erythrocytes (red blood cells), leucocytes (white blood cells) and platelets (cell fragments).

Erythrocytes are adapted to carry oxygen:

- Biconcave disc shape increases surface area for O_2 uptake.

- Cytoplasm is composed of red-pigment haemoglobin, which is specifically adapted for temporary carrying of oxygen.

- Possess no nucleus or organelles and so have more room for haemoglobin.

3 Haemoglobin can either exist in a deoxygenated form or in one of four oxygenated forms:

1	Hb_4 DeoxyHb	$+ O_2 \longleftrightarrow$	Hb_4O_2 OxyHb 1
2	Hb_4O_2 OxyHb 1	$+ O_2 \longleftrightarrow$	Hb_4O_4 OxyHb 2
3	Hb_4O_4 OxyHb 2	$+ O_2 \longleftrightarrow$	Hb_4O_6 OxyHb 3
4	Hb_4O_6 OxyHb 3	$+ O_2 \longleftrightarrow$	Hb_4O_8 OxyHb 4

- O_2 is taken up and released least readily by Hb_4 (deoxyhaemoglobin).
- O_2 is taken up most readily by Hb_4O_6. Similarly, O_2 dissociates most readily from Hb_4O_8.
- Once haemoglobin has bound O_2 it becomes more reactive with each molecule that binds.
- When haemoglobin has bound its full complement of O_2 it is at its most unstable.

4 The characteristics of haemoglobin give rise to the sigmoid (s-shaped) curve for the dissociation of O_2 from haemoglobin.

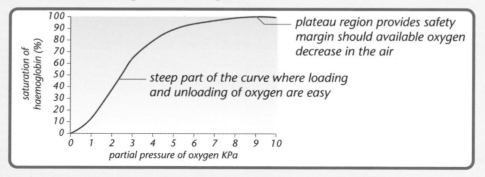

plateau region provides safety margin should available oxygen decrease in the air

steep part of the curve where loading and unloading of oxygen are easy

These characteristics of the uptake of oxygen have several advantages:

- Haemoglobin can become fully saturated with O_2 at relatively low partial pressures, i.e. in the lungs.

- Haemoglobin releases O_2 easily when there is a slight decrease in partial pressure, i.e. at respiring tissue.

5 The oxygen-dissociation curve is affected by slight changes in blood pH. These are caused by changes in the levels of carbonic acid, determined by the volume of CO_2 produced through respiration.

- At decreased pH (more acidic) the curve shifts to the right, as the haemoglobin has a reduced affinity for O_2.
- At increased pH (more alkaline) the curve shifts to the left, as haemoglobin has increased affinity for O_2.

So:

$\uparrow$ **Muscular** $\Rightarrow$ $\uparrow$ **respiration** $\Rightarrow$ $\uparrow$ **CO_2** $\Rightarrow$ $\downarrow$ **pH** $\therefore$ **Dissociation curve**
activity **produced** **shifts to the right.**

This is known as the Bohr effect and means that more O_2 is released to the cells that require it, i.e. those actively respiring.

Bohr effect means the cells with the greatest O_2 demand receive the most.

6 Leucocytes are the white blood cells, which are involved in defence.

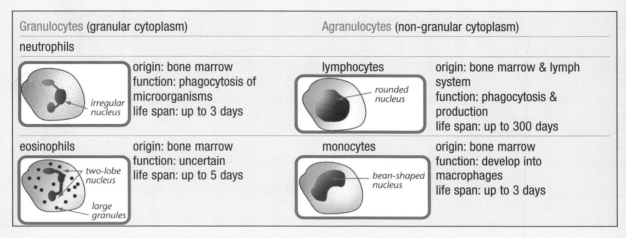

Granulocytes (granular cytoplasm)	Agranulocytes (non-granular cytoplasm)

neutrophils

origin: bone marrow
function: phagocytosis of microorganisms
life span: up to 3 days

irregular nucleus

lymphocytes

rounded nucleus

origin: bone marrow & lymph system
function: phagocytosis & production
life span: up to 300 days

eosinophils

two-lobe nucleus

large granules

origin: bone marrow
function: uncertain
life span: up to 5 days

monocytes

bean-shaped nucleus

origin: bone marrow
function: develop into macrophages
life span: up to 3 days

Phagocytosis is the engulfing and digestion of microorganisms by some leucocytes.

7 Mammals have a double circulatory system. Blood passes through the heart twice before being pumped to the tissues of the body. The circulatory system can be divided into two:

- Pulmonary (including the heart and lungs).
- Systemic (including the heart and rest of body). This means the heart has two sides.

8 The human heart rate is approximately 70 beats/minute. The cardiac cycle can be divided into a series of steps:

Step 1 Heart beat starts in the sinoatrial node (SAN), which is myogenic, i.e. generates rhythmic impulses without stimulation from the brain.

Step 2 Impulses spread across the atria, causing atrial systole, a uniform contraction emptying the atria into the ventricles. A thin layer of non-conductive tissue prevents the impulse reaching the ventricles.

Step 3 The SAN stimulates the atrioventricular node (AVN), which spreads the impulse to the base (apex) of the ventricles via conducting tissues in the Bundle of His.

Step 4 Purkinje fibres then transmit the impulse back up the heart, squeezing the blood out of the ventricles into the arteries.

9 When blood enters the arteries, the highest pressure is termed systolic pressure, and the lowest diastolic pressure, giving two values. Normal blood pressure is approximately 120/80 mm of Hg.

Blood pressure is monitored by sensory cells in the aorta and right atrium. They transmit impulses via the vagus nerve to the cardiac control centre of the Medulla oblongata in the hind brain. Sensory cells in the carotid sinus (in the neck) also relay impulses along the sinus nerve. The cardiac-control centre regulates the SAN through the two parts of the autonomic nervous system (ANS):

- Sympathetic. Secretes noradrenaline which speeds up the heart rate by acting on the SAN and ventricular walls.
- Parasympathetic. Secretes acetyl choline, which slows down the heart rate by acting on either the SAN or vagus nerve.

10 Hydrostatic pressure (due to blood pressure) forces water, ions (glucose, amino acids, salts) and small molecules out of the capillary at the arterial end. This forms tissue fluid, which bathes the tissue cells. Nutrients and oxygen diffuse to the cells through the tissue fluid, and metabolic wastes (carbon dioxide, lactic acid and urea) diffuse from the cells to the plasma.

> *Remember both atria contract together and then both ventricles together.*

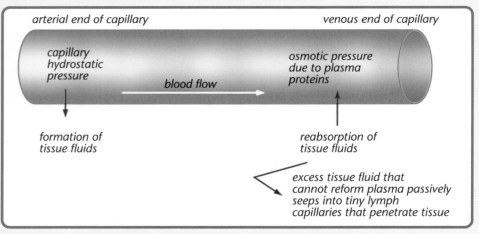

Water is drawn back into the capillary at the venule end, due to the reduced hydrostatic pressure, and inward osmotic pressure is generated by plasma proteins remaining in the capillary.

Have you improved?

1 The cardiac cycle in a human heart has two distinct phases during which the blood flows from the atria to the ventricles and is then pumped either around the body or to the lungs.

(a) (i) Name and describe the two stages of the cardiac cycle.

(ii) Name the structure which initiates electrical activity in the heart

(b) Explain the significance of the following in the cardiac cycle:

(i) Contraction of the ventricles starts at the base of the heart.

(ii) There is a slight delay in the passage of electrical activity between the atria and ventricles.

2 The diagrams below show cross-sections of human blood vessels (not to scale):

(a) Name the three blood vessels labelled I, II and III.

(b) State one way in which each of the vessels is adapted to its function.

(c) Which of the three vessels has a wall allowing movement of white blood cells in and out?

Transport in Plants

How much do you know?

1 Root hair cells create a _____ _____ _____ for the absorption of water and mineral ions.

2 Water can move through the _____ of a root by three pathways: the _____ route travels through the cell walls, the symplastic route travels through the _____ and the _____ travels through the vacuoles.

3 Water and minerals are transported in _____ vessels; _____ and amino acids are transported in phloem.

4 Three main forces move water through xylem: (1) _____ pressure, (2) _____ tension, and (3) transpiration.

5 _____ is the movement of mineral and organic compounds through the phloem, possibly according to the _____ hypothesis.

6 _____ are plants adapted to living in dry conditions. They have _____ stomata to reduce water loss by _____.

7 _____ are plants adapted to living in or on open water.

8 Maize can grow at a fast rate due to the presence of _____ _____ cells.

9 Insecticides can increase in concentration along a food chain and this is called _____. It becomes _____ to top consumers.

DAY

5

Answers

9 biomagnification, toxic
5 translocation, mass-flow 6 xerophytes, sunken, transpiration 7 hydrophytes 8 bundle sheath
1 large surface area 2 cortex, apoplastic, cytoplasm, vacuolar 3 xylem, sucrose 4 root, cohesion

If you got them all right, skip to page 59

Learn the key facts

1 The structure of a root is shown in the diagram below:

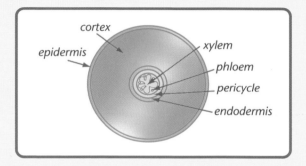

Root hair cells create a large surface area for the absorption of water. Cortex consists of parenchyma cells which are unspecialised cells.

Endodermis is a single layer of cells which have a layer of suberin called the Casparian strip in the cell walls. Water flowing in the apoplastic pathway has to move into the symplastic pathway due to the Casparian strip.

2 Water moves into the roots by osmosis at root hair cells. Mineral ions are taken up by active transport against the concentration gradient, e.g. magnesium ions are needed to produce chlorophyll and a lack of magnesium causes chlorosis.

Water travels across the cortex of the root by three different pathways: the symplastic route – through the cytoplasm; the apoplastic route – through the cell walls; and the vacuolar route – through the vacuoles.

DAY

5

3

Tissue	Characteristics	Structure & function
Xylem vessels Carry water and minerals e.g. nitrate	• Made of xylem vessel elements and tracheids • Have thick, heavily lignified secondary cell walls and a hollow central lumen	• Dead, lignified cells provide support to the plant
Phloem vessels Carry sucrose, amino acids, fatty acids, glycerol and hormones	• Sieve tubes (cylindrical column of cells) with end walls perforated forming sieve plates • Companion cells connected to sieve tubes via numerous plasmodesmata	• Companion cells may regulate activity of sieve tube • All cells are living allowing control of transport

spiral vessel annular vessel reticular vessel pitted vessel

movement of water

companion cell

sieve tube element

plasmodesmata

4 Three main forces move water and minerals up the xylem:

- Root pressure. Mineral ions (e.g. K^+) are actively pumped into the pericycle cells around the xylem from the adjacent endodermal cells. This moves water into the xylem via osmosis and creates an upward force.
- Cohesion:tension theory. H_2O molecules are polar and so both attract each other (cohesion) and are attracted to the walls of the xylem (adhesion). These forces can draw water up even very tall trees.
- Transpiration. H_2O molecules evaporate from the spongy mesophyll cells in the leaves and diffuse out via the stomata. This lost H_2O causes H_2O from adjacent cells to diffuse in, drawing H_2O from the xylem. The cohesion of H_2O molecules draws water up from the roots.

Transport in Plants

Various factors will increase the rate of transpiration:

- Increasing light intensity causing full stomata opening.
- Increasing temperature increases kinetic energy of H_2O molecules, causing faster movement.
- Decreasing humidity means air can hold more water and so increases the concentration gradient between air and leaves.
- Increased air movement removes saturated air from around leaves.
- Decreased stomata density prevents overlap of diffusion shells.

5 Translocation is the bi-directional movement of mineral and organic compounds manufactured in the plant, e.g. sucrose in the phloem.

The mass-flow hypothesis says that translocation is driven by glucose in the leaves raising their osmotic pressure, causing water to enter. This generates a turgor pressure, causing movement along the phloem. At the roots these sugars are removed and water is lost, returning to the leaves.

There are various limitations with the mass-flow hypothesis:

- Sieve plates would resist the turgor pressure.
- Phloem transport is bi-directional.
- It does not explain why sieve tubes and companion cells are metabolically active, nor the role of ATP in translocation.

6 Xerophytes are plants adapted to living in dry environments. Their features for survival include:

- Sunken stomata – on inside to reduce water loss.
- Thick waxy cuticle – prevents loss of water.
- Rolled leaves – trap moist air (e.g. *Ammophila* – marram grass).
- Trichomes (hairs) – trap moist air.
- Spines – reduced surface area of leaf. Therefore, less water loss by transpiration.
- Extensive root system – shallow, wide root system to absorb dew and light rain, long roots to absorb water deep underground.
- Diurnal closing of stomata – stomata close at midday to reduce water lost by transpiration.

Rate of transpiration is dependent on the size of the diffusion gradients.

Phloem tissue is living cells, unlike xylem.

DAY

5

7 Hydrophytes are plants which are adapted to living in or on open water. Their features include:

- No waxy cuticle – evaporation of water is not a concern.
- Little or no xylem tissue – particularly no tracheids needed for support.
- Air spaces in stems and leaves – contain carbon dioxide and oxygen and help buoyancy.

8 Cultivated plants and cereals are adapted structurally and physiologically to survive in different environments. Humans can manipulate the environment to increase productivity. For example:

(a) Tropical plants or C_4 plants (e.g. maize) are adapted to living in hot climates. They have cells called bundle sheath cells, which receive malate from the mesophyll cells and this is broken down into pyruvate and carbon dioxide. This excess carbon dioxide is used for increased rate of photosynthesis and results in rapid growth.

(b) Rice grows in swamps and is able to withstand high levels of ethanol produced by anaerobic respiration.

9 Pesticides are chemicals used to kill pests.

	Advantages	Disadvantages
Herbicides	Kill weeds that compete with the crop for sunlight, minerals and water (i.e. reduce interspecific competition)	Easily leached into nearby water, killing plants and disrupting food chains
Insecticides	Specifically kill harmful insects, thereby crop yield	Bioaccumulation and biomagnification (i.e. become increasing toxic to top consumers)

45 mins

Time Yourself

Have you improved?

1 The transport system in flowering plants involves both xylem and phloem tissue.

(a) Name the two types of cells characteristic of phloem tissue.

(b) In an experiment, an aphid was allowed to feed on a plant stem. The experimenter then anaesthetised the aphid and cut off its proboscis (feeding apparatus). The exudate produced was then collected and analysed.

> Aphids are parasites on plants.

(i) Suggest which tissue the exudate is being produced from.

(ii) State two substances that you would expect to find in the exudate.

(iii) State one substance that will not be present.

2 The diagram below shows a cross section through marram grass (*Ammophila arenaria*).

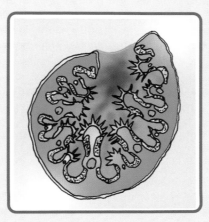

State three xeromorphic adaptations and explain how this reduces the rate of transpiration.

DAY

5

Reproduction

How much do you know?

1 Reproduction may be sexual or _____. Sexual reproduction involves _____ parents whereas asexual reproduction always involves one parent.

2 Pollination is the transfer of _____ from the _____ to the stigma. Insect pollinated flowers have large _____ petals, which are usually scented.

3 At fertilization, one pollen nucleus fuses with the _____ _____ producing a zygote which develops into the _____. The second pollen nucleus fuses with the diploid cell, producing a _____ nucleus which develops into _____ tissue.

4 Cross-pollination can be prevented by the anthers maturing before the stigmas, called _____, or dioecious, which ensures separation of _____.

5 Spermatogenesis is the production of spermatozoa and is triggered by the hormone _____.

6 At puberty, females release one egg cell – a secondary _____ – from alternate ovaries per _____.

7 FSH stimulates the development of a _____ _____ in an ovary and stimulates it to release oestrogen. In anticipation of fertilization, oestrogen promotes _____ of the endometrium.

8 _____ is the ejaculation of sperm into the female vagina, controlled by the _____ nervous system.

9 The placenta has two main functions: exchange of material between _____ and _____ blood; and _____ production.

10 Birth is triggered by oxytocin from the _____ _____ gland and prostaglandins from the _____.

Answers

1 asexual, two   **2** pollen, anther, coloured   **3** egg cell, embryo, triploid, endosperm
4 protandry, gametes   **5** FSH   **6** oocyte, ovulation   **7** Graafian Follicle, thickening/vascularisation
8 copulation, sympathetic   **9** foetal, maternal, hormone   **10** anterior pituitary, placenta

If you got them all right, skip to page 66

Reproduction

Spend no more than **45** mins on this topic

Learn the key facts

1 Reproduction may be sexual or asexual.

Asexual: e.g. bacteria	Sexual: e.g. animals and higher plants
One parent only	Usually two parents
No gametes produced	Gametes produced: small, mobile, male gametes and large, stationary female gametes e.g. sperm, pollen and ovas
Mitosis only involved	Mitosis and meiosis involved. Meiosis usually produces haploid gametes
Offspring identical to parents and each other, except when mutations occur	Offspring varied from parents and each other because of mutations, crossing over and independent assortment in meiosis, combination of parental genomes and random fertilization. Variation allows evolution
Usually many offspring, produced very rapidly	Usually fewer offspring, produced more slowly

Sexual reproduction always produces offspring which are different from the parents. This variation allows the process of natural selection to operate and hence organisms evolve.

2 In flowering plants, pollen is produced in the stamens, which each consist of an anther and a long filament. The transfer of pollen from the anther to the stigma is called pollination. There are many differences between wind- and insect-pollinated flowers.

You must be able to label the structure of a flower.

DAY 6

61

Wind-pollinated flower	Insect-pollinated flower
Small petals (usually green) or petals absent; flowers inconspicuous	Large coloured petals; conspicuous flowers/inflorescences
Not scented	Scented
Nectaries absent	Nectaries present
Large branched and feathery stigma hanging outside flower Pendulous stamens hanging outside flower to release pollen	Small sticky stigma to hold pollen, usually enclosed within flower Stamens enclosed within flower
Large quantities of pollen Pollen grains relatively light and small; dry, often smooth walls	Less pollen produced Pollen grains relatively heavy, large, possessing spines or being sticky for attachment to insect body
Simple flower structure	Complex flower structure

3 After being deposited on the stigma, the pollen grain produces a pollen tube which goes down the style and eventually enters the ovule via a tiny hole called the micropyle. The generative nucleus of the pollen grain forms two male nuclei. These pass down through the pollen tube to the female egg cell. One of the male nuclei fuses with the egg cell, producing a zygote which develops into the embryo. The second male nuclei fuses with the diploid central cell, producing a triploid nucleus which develops into endosperm tissue, which provides food for the developing embryo.

4 Many plant species prevent self-pollination (fertilisation of female gametes with male gametes from the same individual). Mechanisms for ensuring cross-pollination include:

- Dichogomy – anthers mature before stigmas (protandry) and are then shed, e.g. dead nettles, or stigmas mature before anthers and are then shed (protogyny).
- Dioecious plants – separate male and female plants, ensuring separation of gametes.

You must be able to label the male and female reproductive systems.

Self-fertilisation can increase the risks from harmful alleles in a population.

5 In mammals, spermatogenesis is the production of spermatozoa. It occurs in the seminiferous tubules of the testes and is triggered by FSH from the anterior pituitary region of the brain.

			chromosome number
mitosis	germinal epithelial cell	(2n)	46
mitosis	spermatogonium	(2n)	46
meiosis I	primary spermatocyte	(2n)	46
meiosis II	secondary spermatocyte	(n)	23
maturity	spermatids	(n)	23
	spermatozoa	(n)	23

6 Oogenesis is the production of egg cells and occurs in the ovaries of the female foetus. Oogenesis follows similar stages to spermatogenesis. At puberty, one egg cell – a secondary oocyte – is released per month in ovulation. This egg cell has undergone the first meiosis division but has stopped at metaphase of the second meiotic division. This is completed only if the egg cell is fertilized by a sperm.

The endometrium is preparing for possible implantation of a fertilised oocyte.

7 The female menstrual cycle is controlled by hormonal secretion.

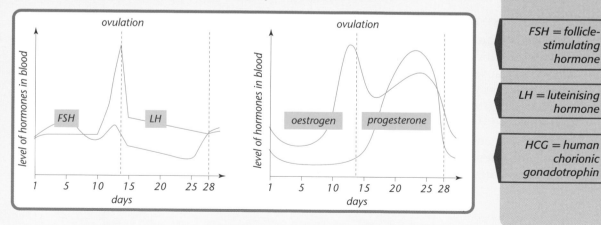

FSH = follicle-stimulating hormone

LH = luteinising hormone

HCG = human chorionic gonadotrophin

- Day 2   Increasing FSH from the anterior pituitary stimulates the development of a Graafian Follicle in an ovary and stimulates the follicle to release oestrogen.
- Day 12   Oestrogen causes the thickening and vascularisation of the endometrium of the uterus, increases LH from the anterior pituitary, and decreases FSH (negative feedback).
- Day 14   Increasing LH leads to ovulation (release of secondary oocyte) from Graafian Follicle.

  Increasing LH and prolactin causes the Graafian Follicle to develop into the corpus luteum.

  The corpus luteum produces progesterone and oestrogen which maintain the endometrium.

  Increasing progesterone leads to a decrease in release of FSH and LH, so no more follicles develop.

- If fertilization occurs the zygote divides mitotically to form a blastocyst, which implants in the endometrium.

The blastocyst secretes HCG, which maintains the corpus luteum and therefore increases progesterone and oestrogen.

- Week 10   The placenta forms and produces progesterone.
- If no fertilization occurs, the corpus luteum stops producing progesterone and oestrogen. FSH production, which had previously been inhibited by high progesterone, now increases again and a new follicle begins to develop. Decreasing progesterone and oestrogen causes the endometrium to disintegrate and it is lost from the body with blood, in the menstrual flow.

8 Copulation is the ejaculation of sperm by the male into the vagina of the female. Increased blood flow causes the penis to become erect (controlled by the parasympathetic nervous system) and to and fro movements of the erect penis within the vagina cause the release of sperm by peristalsis (controlled by the sympathetic nervous system). The sperm swim, using flagella, into the oviducts. To enable fertilization the acrosome in the head of the sperm bursts, releasing enzymes which digest the outer layer of the egg. As soon as one sperm has penetrated the egg a fertilization membrane forms, preventing any other sperm entering. The egg now completes a second meiotic division. The sperm and egg nuclei fuse to form a diploid zygote.

9 The zygote divides by mitosis to form a blastocyst (hollow ball of cells), which burrows into the endometrium of the uterus. The outer trophoblastic cells develop into the disc-shaped placenta.

The placenta has two main functions:

- Provides a large surface area (due to choriomic villi and microvilli) for exchange of material between maternal and foetal blood, via diffusion.

From foetus to mother	From mother to foetus.
Waste: CO_2 and urea	Nutrients, e.g. glucose, O_2
	Antibodies (give passive immunity)
	Drugs, toxins and pathogens

- Hormone production.
  Human chorionic gonadotropin maintains corpus luteum.
  Progesterone and oestrogen maintain endometrium and relax uterus muscles.

10 Birth is triggered by:

- Decreasing progesterone, which had previously inhibited uterine contractions.
- Oxytocin from the anterior pituitary and prostaglandins from the placenta which stimulate contractions.

Lactation involves the production of milk for the baby and is stimulated by prolactin from the pituitary gland.

Placenta prevents mixing of maternal and foetal blood.

1

2

3

4

5

DAY

6

7

Have you improved?

1 The diagram below shows some of the stages of the development of pollen grains:

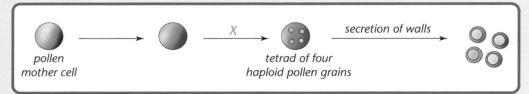

pollen
mother cell

X

tetrad of four
haploid pollen grains

secretion of walls

(a) Name process X.
(b) Why is it important that gametes are haploid?

2 Aspen trees may reproduce asexually. Suggest explanations for the following observations:
(a) Following land clearance, aspen forests may colonise very rapidly.
(b) Fungal diseases may spread extremely rapidly from one individual tree to another.

3 The diagram below shows the structure of a mature sperm.

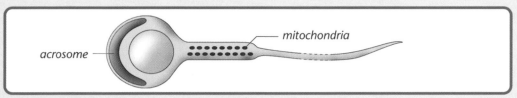

mitochondria

acrosome

(a) What is the function of the acrosome?
(b) Suggest an explanation for the large number of mitochondria in the middle piece.

4 Suggest explanations for the following observations:
(a) Pregnancy kits test for the hormone HCG.
(b) Contraceptive pills often contain a mixture of progesterone and oestrogen.

How much do you know?

1 Parasitic nutrition is a close association between two organisms which is _____ to the parasite and harmful to the _____. In contrast, _____ nutrition is beneficial to both organisms.

2 A _____ is the place where an organism lives in an ecosystem and the _____ _____ is the role an organism plays in an ecosystem.

3 In food chains each stage is known as a feeding or _____ level. Transfer of energy between levels is inefficient because energy is _____ at each trophic level.

4 Pyramids of numbers represent the numbers of _____ at each trophic level. Their main disadvantage is that they can be _____, e.g. when based on a single tree.

5 _____ convert dead organic matter into small particles, which aids decomposers to break down large organic molecules _____.

6 In ecosystems, _____ primary productivity is the energy remaining after energy loss via _____ from the plant.

7 Fermentation of biomass by _____ bacteria produces biogas, which is mainly _____. The only pollutant produced is _____.

8 _____ is the clearance of woodland, often to create land for crops. The consequences include increased _____ erosion, _____ of rivers and habitat loss.

9 Stratospheric ozone absorbs _____ radiation from the sun. Depletion has occurred during the twentieth century due to release of _____ from aerosols and as coolants.

10 Acidic _____ is the deposition of acidic gases from the atmosphere, produced by combustion of _____ fuels. The consequences include _____ of freshwater.

11 Greenhouse gases allow entry of _____ radiation from the Sun, and _____ loss of re-radiated long-wave radiation.

Answers

1 beneficial, host, mutualistic 2 habitat, ecological niche 3 trophic, lost 4 organisms, inverted 5 detritivores, chemically 6 net, respiration 7 anaerobic, methane, CO_2 8 deforestation, soil, silting 9 short-wave, CFCs 10 precipitation, fossil, acidification 11 short-wave, delay

If you got them all right, skip to page 73

Ecology

Learn the key facts

1 Living organisms obtain organic molecules in a variety of ways:

Autotrophic nutrition, e.g. photosynthesis, involves conversion of inorganic raw materials, e.g. CO_2 and H_2O, into complex organic molecules, e.g. carbohydrates. This process requires energy, such as from sunlight.

Heterotrophic nutrition is the feeding on complex ready-made organic food obtained in the diet. This can be divided into:

a) Holozoic nutrition – taking food into the body, where it is digested, absorbed and assimilated.

b) Saprobionic nutrition – feeding on dead or decaying organic matter, as performed by *Rhizopus* bacteria.

c) Parasitic nutrition – close association between two organisms of different species, beneficial to one (the parasite) and harmful to the other (the host). Parasites obtains food and usually shelter from the host, e.g. *Taenia* (tapeworm) found in the human gut.

d) Mutualistic nutrition – close association between two living organisms, beneficial to both, e.g. *Rhizobium* bacteria living in root nodules of leguminous plants.

> *You must learn these definitions.*

2 An ecosystem is any community of organisms interacting with one another, and with their physical environment. Within an ecosystem, organisms live in a particular place – habitat – defined by its physical characteristics. The ecological niche is the role played by a species in the ecosystem. This includes its habitat, plus its position in the community, e.g. producer, predator.

> *Ecological niche is the 'job' an organism has in an ecosystem.*

All ecosystems are composed of:

• Producers – autotrophic organisms, mainly green plants and algae, providing a source of organic molecules for other organisms in the ecosystem.

• Consumers – primary consumers (herbivores) feed on producers, and secondary consumers (carnivores) feed on herbivores.

3 Food chains are linear sequences of organisms in a feeding relationship. Each stage is known as a feeding or trophic level.

Oak tree	→	Earthworms	→	Shrews	→	Owls
Producer	→	Primary consumer	→	Secondary consumer	→	Tertiary consumer
Producer	→	Herbivore	→	Carnivore	→	Carnivore

Arrows indicate energy flow between trophic levels, starting at the producer.

Food webs are a system of numerous food chains and are a more realistic representation of feeding relationships, since many organisms feed on more than one species and at more than one trophic level.

There are rarely more than four trophic levels because there is not enough energy to support more. The transfer of energy between trophic levels is inefficient, with energy lost at each trophic level for several reasons:

- Not all of the organisms at any trophic level are eaten.
- Some organic molecules, e.g. cellulose in plants, are undigestable to some organisms and so not available to the next trophic level.
- Energy is lost at each trophic level as heat due to respiration.

4 Pyramids of numbers, biomass and energy are used to measure the productivity of organisms in an ecosystem.

Pyramid of numbers	Pyramid of biomass	Pyramid of energy
number of individuals	*dry mass/g m^{-2}*	*kJ m^{-2} y^{-1}*
Represents number of organisms at trophic level	Total dry mass of organic matter per trophic level (in a unit area or volume)	Organisms are converted to their energy equivalent kJ m^{-2}yr^{-1}
Advantages		
• Easy to produce • Non-invasive (no organisms killed)	• More realistic representation of food web as usually upright	• Always upright, because energy is always lost at each trophic level
Disadvantages		
• Can be inverted, e.g. one oak tree at base • Numerous small consumers, e.g. parasites, unbalance the pyramid	• Needs more data to construct because it is based on **dry** mass • Organisms are killed • Can be inverted due to seasonal variations, e.g. variations in phytoplankton populations	• Difficult to construct. Many samples must be taken over a long period and all samples must be killed, dried, weighed and burnt to measure energy content

In the pyramid diagram, the pyramid of numbers shows levels labelled *tertiary*, *secondary*, *primary consumers*, *producers*. The pyramid of energy shows values 90, 1500, 14000, 88000.

Ecology

5 Detritivores are animals that feed on dead organic matter mechanically breaking down (comminuting) large particles into small ones. Decomposers are bacteria and fungi that feed on dead organisms chemically breaking down (digesting) large organic molecules into small ones.

Detritivores and decomposers form the detrital food chain. This is a vital part of any ecosystem as dead organic material is broken down into its simple inorganic ions making these mineral nutrients available for uptake by plants, i.e. inorganic ions are recycled.

6 Productivity is defined as the rate at which organic matter or energy is assimilated (built up) into new tissues, per unit area, per unit time. Gross primary productivity (GPP) is the total amount of organic matter or energy fixed by green plants. Net primary productivity (NPP) is the amount of energy remaining after energy is used in respiration, and therefore available to herbivores.

This relationship can be expressed using a simple equation:

Gross Primary Productivity (GPP) − Respiration (R) = Net Primary Productivity (NPP)

7 Humans utilise energy resources to produce electricity and for transport:

Energy resource	Key Points
Non-renewable	Found in finite supply and will eventually run out
Coal and oil	• Account for most energy use by humans • Pollution problems include: acid rain, CO_2 causing the enhanced greenhouse effect
Renewable	Infinite supply, so will not run out if managed sustainably
Biomass e.g. wood and straw	• Energy extracted by direct burning or in a boiler to produce steam to drive an electricity turbine • Significantly less pollution than coal and oil • Growth absorbs CO_2, so reducing enhanced greenhouse effect
Biogas	• Fermentation of biomass (e.g. dung) by bacteria, producing methane • Occurs in a digester (enclosed tank) at around 30–40°C using methanogenesis bacteria • Burning only produces CO_2 as a pollutant

Action of detritivores increases the surface area for decomposers.

8 Deforestation is the clearance of woodland, often to create land for crop fields, housing or mining and to supply wood for fuel. The extent of damage has increased with increasing mechanisation and growing human populations. The effects of deforestation include:

- Increased soil erosion and leaching of nutrients as rain impact and run-off increases, along with wind erosion.
- Rivers become silted up with eroded material and may flood due to the increased volume of water reaching them.
- Habitats are lost for many species, so reducing species diversity.

9 Stratospheric ozone (O_3) found between 15 km and 30 km, absorbs short-wave radiation from the Sun, protecting living organisms from this radiation which can cause skin cancers and DNA mutation.

Human activity has caused significant thinning of the O_3 layer, particularly at the poles, mainly due to the use of CFCs (in aerosols and as coolants) and nitrous oxides (released from artificial fertilizers). When released into the atmosphere these chemicals break down O_3 faster than it is reformed. This has led to increased cases of skin cancer and reduced crop yield due to damage to vegetation, particularly in the southern hemisphere where greatest thinning has occurred.

10 Acidic precipitation is the deposition of acidic gases (pH 4–4.5) from the atmosphere produced by combustion of coal and oil (fossil fuels). The principal gases are sulphur dioxide and nitrogen dioxide.

Gases may be deposited directly onto surfaces (dry deposition) or dissolved in water droplets (wet deposition), which causes varied effects:

- Decline in forests due to acid soil conditions and by making heavy metal ions in the soil available for uptake by roots, so poisoning them.
- Acidification of freshwater, killing fish and invertebrates.
- Chemical weathering of buildings and monuments.

11 The greenhouse effect is the natural warming of the Earth's atmosphere due to greenhouse gases, e.g. CO_2. These gases allow entry of short-wave radiation from the Sun, but delay the loss of re-radiated long-wave radiation, causing an overall warming effect.

Human activities have increased the concentration of greenhouse gases in the atmosphere. This enhanced greenhouse effect has been linked with global climate change, including increased temperatures and declining rainfall.

Burning of fossil fuels, e.g. at power stations, releases greenhouse gases.

Have you improved?

1 Define the following ecological terms:

(a) Community

(b) Pyramid of energy

(c) Succession.

2 Energy flow through an ecosystem is linear.

(a) Explain what is meant by this.

(b) Explain why most food chains are limited to four trophic levels.

3 The pyramid of biomass represents the dry mass (g/m³) of plant plankton and animal plankton in the North Atlantic.

zooplankton
phytoplankton

Comment on the shape of this pyramid.

4 Describe the cause and effect of each of the following:

(a) Greenhouse effect

(b) Ozone depletion

(c) Acid precipitation.

Exam Practice Questions

1 Put a tick in the box if the statement is correct and a cross if it is incorrect. (4)

Carbohydrate	Insoluble	Forms fibrils	Reducing sugar	Polymer of α-glucose
Starch				
Cellulose				
Maltose				
Glycogen				

2 The diagram below shows the structure of a cell from the ileum in the small intestine.

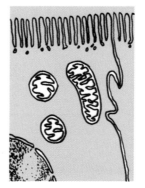

(a) Name the type of cell shown. _____ (1)

(b) Describe and explain two visible features shown that are useful in the uptake of food molecules. (4)

(i) _____

_____

(ii) _____

_____

(c)   Define a tissue. _____   (2)

3 Fill in the gaps in the following passage:

During inspiration, the _____ intercostal muscles contract, causing the rib cage
to move _____ and _____. The _____ muscle contracts causing it
to become _____. This _____ the _____ of the thorax causing a
_____ in pressure and therefore _____ rushes into the lungs.   (9)

4 Complete the table below.   (8)

Vitamin	Solubility	Function	Deficiency
	fat	vision	
C			scurvy
		absorbs calcium	rickets
B1		release of energy	

5 The following drawing shows a transverse section through a root.

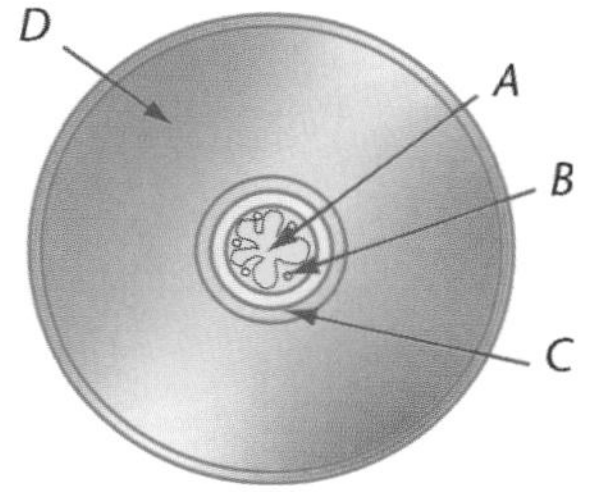

(a) Label the parts.

A _____   C _____

B _____   D _____   (4)

(b) Describe how water moves into a root and across a root to the xylem tissue. (3)

(c) How is the structure of the xylem tissue adapted for its function? (2)

6 The graph below shows the oxygen dissociation curve for human haemoglobin.

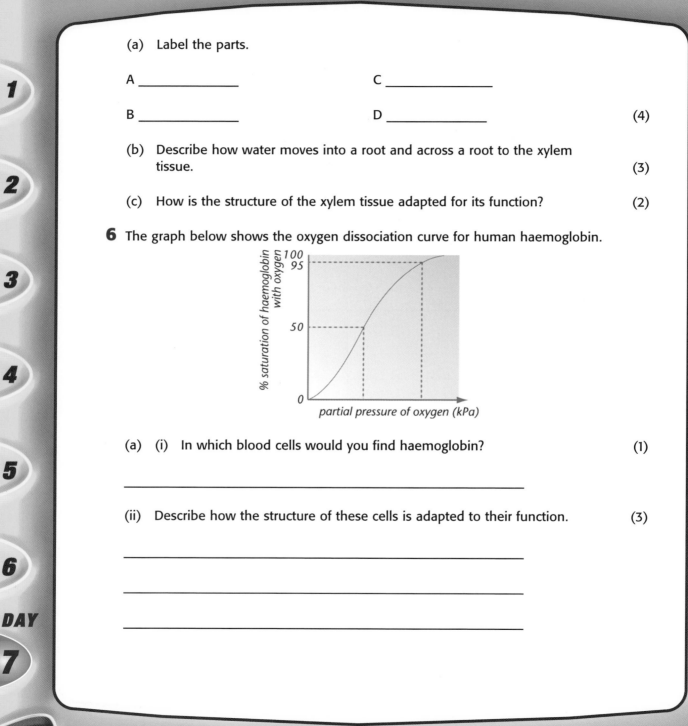

(a) (i)   In which blood cells would you find haemoglobin? (1)

_____

(ii) Describe how the structure of these cells is adapted to their function. (3)

_____

_____

_____

(b) (i) Using the information in the graph and your own knowledge, explain how the haemoglobin is able to take up oxygen in the lungs and unload in the tissues. (5)

_____

_____

_____

_____

_____

(ii) What would be the effect on the partial pressure for a 50% saturation of an increase in the partial pressure of CO_2? (1)

_____

(iii) Explain how this may be an advantage in supplying oxygen. (2)

_____

_____

7 (a) Explain how the shape of an enzyme molecule contributes to its specificity. (3)

_____

_____

_____

(b)   When referring to enzyme action, define the following terms:

(i)   The induced fit hypothesis   (2)

_____

_____

(ii)   Competitive inhibition   (2)

_____

_____

(c)   An experiment was conducted to determine the effect of pH on the activity of the enzyme catalase (found in potato discs) when supplied with its substrate hydrogen peroxide.

(i)   Suggest what you could use to measure the rate of reaction.   (2)

_____

_____

(ii)   State three factors that you must ensure to achieve accurate results.   (3)

_____

_____

_____

8 (a)   Define the following ecological terms:

(i)   Community   (2)

_____

_____

DAY

1

2

3

4

5

6

7

(ii)   Pyramid of biomass   (3)

_____

_____

(iii) A population   (2)

_____

_____

(b)   The pyramids below were calculated for a food chain in an oak woodland.

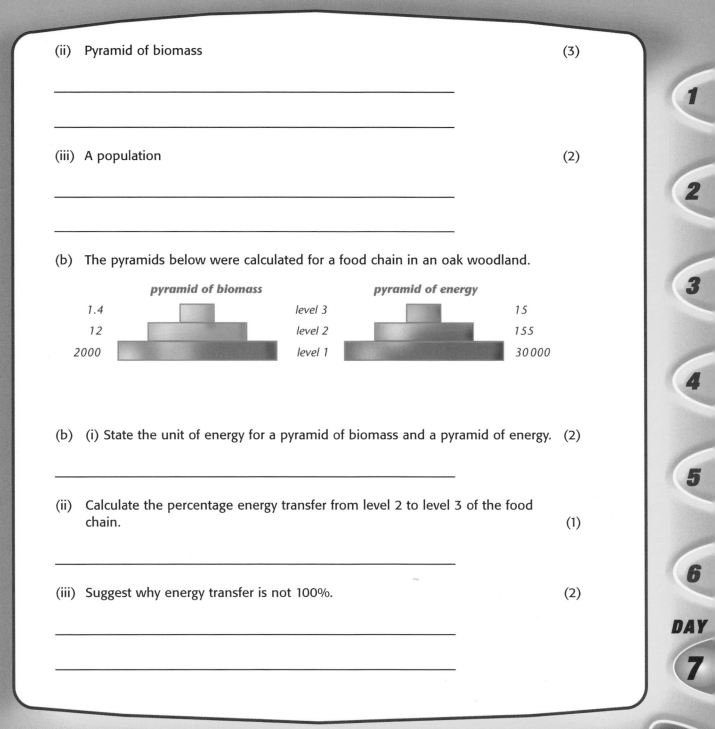

pyramid of biomass		*pyramid of energy*	
1.4		level 3	15
12		level 2	155
2000		level 1	30 000

(b)   (i) State the unit of energy for a pyramid of biomass and a pyramid of energy.   (2)

_____

(ii)   Calculate the percentage energy transfer from level 2 to level 3 of the food
       chain.   (1)

_____

(iii) Suggest why energy transfer is not 100%.   (2)

_____

_____

9 The diagram below shows the generalised structure of a flower.

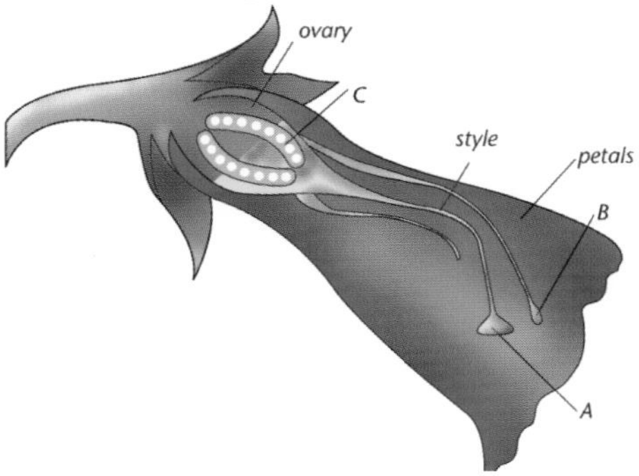

(a) (i) Name and give one function of structures A, B and C.   (6)

A_____

Function_____

B_____

Function_____

C_____

Function_____

(ii) Identify and explain the type of pollination for this flower structure.   (3)

_____

_____

_____

(b) The oestrogen or menstrual cycle in mammals is controlled by hormones secreted from two sites.

(i) Identify these two sites. (2)

_____

(ii) Define the term hormone. (2)

_____

_____

(c) Describe the role of each of the following hormones in the oestrogen cycle:

(i) Progesterone (2)

_____

_____

(ii) FSH (2)

_____

_____

10 Read through the following passage and complete the text by filling in the blank spaces with the most appropriate biological term. (12)

A molecule of DNA is composed of individual monomers called _____. Each monomer is composed of three units, a nitrogenous base, joined to a _____ sugar and a _____ group. RNA exists in three different forms in cells. Ribosomal RNA is found in high concentrations in the _____ region of the nucleus. _____ RNA passes from the nucleus to the cytoplasm via pores in the _____ _____. In the cytoplasm, this binds with _____ RNA, in the presence of ribosomes, which bring in _____ _____ to produce a polypeptide chain. The two stages of protein synthesis are known as _____ and _____.

Answers on page 86

Cells, Tissues and Organisms

1

	Simple diffusion	Facilitated diffusion	Active transport
Is ATP required?	N	N	Y
Rate of movement	Slow	Fast	Fast
Direction of transport in relation to concentration gradient	Along	Along	Against

2 (a)   X = phospholipid   Y = Glycocalyx.
    (b)   Lipids/proteins can move laterally/exchange places.
    (c)   (i)   Synthesis of lipids.
           (ii)   Very active/needs lots of ATP.

Organic Molecules

1 (a)   Must be part of diet; not synthesised by the body.
    (b)   Phosphate group present; only two fatty acids.
    (c)   Carbon; hydrogen; oxygen.

2 (a)   Amino acids.
    (b)   Monosaccharides.
    (c)   Monosaccharides.
    (d)   Fatty acids.

3 (a)   (i)   Condensation; water.
           (ii)   Peptide.
           (iii)   Dipeptide.
    (b)   Sulphur; nitrogen or phosphorus.
    (c)   Sulphur forms disulphide bridges; may be hydrophobic/hydrophilic in tertiary structure; may be acidic/basic/buffer; may be charged allowing electrostatic interaction; capable of hydrogen bonding.

Enzymes and Biotechnology

1 (a)   Active site; steric relationship with substrate; other molecules unable to bind.
    (b)   *Three of*: temperature; enzyme concentration; substrate concentration; presence/absence inhibitor; pH.
    (c)   (i)   Site where substrate binds; forming steric relationship with substrate.
           (ii)   Inhibitor structurally similar to substrate; competes with active site; reversible.

2 (a)   Maintain constant pH.

(b)   Break open cells to release enzyme/catalase; increase surface area.

(c)   Hydrogen peroxide.

(d)   Water bath, maintain suitable temperature range; top-pan balance, constant mass of potato tissue; thermometer, monitor constant temperature.

3 (a)   (i) Pectinase; clarification of fruit juice *or* proteases; biological detergents/stain removal.

(ii) (Catalyse) reactions at normal temperatures/pressures; reduced cost of producing high temperatures and pressures.

(b)   (i) Attaching or trapping an enzyme to an inert, solid support.

(ii) Advantages – reusable, do not contaminate the product, less likely to be denatured by pH/temperature. Disadvantages – high initial cost; cannot be used for complex reactions.

Respiration and Gaseous Exchange

1 (a)   Small organisms have large ratios of surface area to volume for rapid diffusion and short diffusion distance; gases can enter and leave along diffusion gradient. Mammals have small ratio of surface area to volume; long diffusion pathways; skin is impermeable; need internal exchange mechanism that is moist and has large surface area; need ventilation mechanism.

(b)   *Three of*: moist; large surface area; transport mechanism; thin surface.

2 (a)   Diffusion from atmosphere; through stomata into leaf air spaces; dissolves in water in moist cell walls; CO_2 diffuses into cells in solution down diffusion gradient.

(b)   *Two of: l*enticels; ruptured areas of bark; roots; root-hair cells with large surface area.

3 (a)   Inhalation brings oxygen down trachea and down bronchus; through bronchioles; into alveoli.

(b)   Prevents collapse of bronchi when pressure falls during inhalation.

4 (a)   One cell thick; short diffusion path; high rate of diffusion.

(b)   Allows O_2 to dissolve; enables diffusion.

Have you improved: Answers

Protein Synthesis and Nuclear Division

1 (a)   (i)   A = ribosome; B = tRNA.

          (ii)   Peptide bond formation; ATP; ligase enzyme; condensation reaction/loss of water.

   (b)   (i)   UAG.

          (ii)   AGC.

2 (a)   Bivalents/homologous pairs; chiasmata formation/crossing over.

   (b)   (i)   46;

          (ii)   23;

          (iii) 46.

   (c)   Homologous chromosomes pair up/formation of bivalents/synapsis; chiasmata formation/crossing over; genetic material exchange between homologous pair; independent assortment; during metaphase I.

   (d)   *Three of*: independent assortment; variation; production of haploid cells; synapsis occurs/ bivalents; two divisions.

Nutrition

1 (a)   A fungus produces extracellular enzymes called amylases, which digest starch into maltose. This will produce a negative starch test result (it remains yellow-brown).

   (b)   Fungus C

   (c)   Set up a petri dish with starch agar and no fungi – no clearance zones should be present (i.e. this confirms that the fungus causes the clearance zones).

2 (a)   (i)   Stomach.

          (ii)   Duodenum.

          (iii) Wall of the ileum.

   (b)   *Three of*: villi/microvilli; large surface area for absorption; lacteal; transport of fats; rich capillary network/maintain concentration gradient; crypt of Lieberkuhn/Brunner's glands secrete enzymes; smooth muscle/ peristalsis.

   (c)   Peristalsis; mixing/movement of food along the gut.

Transport in Mammals

1 (a)   (i)   Diastole: ventricles filling/atria contracting; systole: ventricles contracting.

          (ii)   SAN/sinoatrial node.

   (b)   (i)   Forces blood up to arteries/arteries at top of ventricles.

          (ii)   Allows atria to empty/ventricles to fill.

2 (a)   I = artery; II = vein; III = capillary.

   (b)   I has muscular wall, narrow lumen; II has valves; III is one cell thick, has pores, narrow lumen.

   (c)   Capillaries.

Transport in Plants

1 (a)   Sieve tube elements; companion cells.

   (b)   (i)   Phloem.

          (ii)   *Two of*: sucrose; amino acids; auxin.

          (iii)   Nitrate.

2 *Three of*: sunken stomata – reduces water lost by transpiration; thick waxy cuticle – waterproof layer reduces water loss; rolled leaves – creates a humid environment and reduces water gradient; trichomes – to trap moist air.

Reproduction

1 (a)   Meiosis.

   (b)   Maintains ploidy; the amount of genetic material would increase at each generation; ensures homologues.

2 (a)   Vegetative reproduction is faster than sexual reproduction; the stems develop from underground root systems.

   (b)   All root systems may be connected; the fungus may be systemic; all trees are genetically identical.

3 (a)   Contains hydrolytic enzymes to digest the outer layer of oocyte; allows penetration of egg.

   (b)   Synthesise ATP to provide energy for swimming.

4 (a)   Blastocyst present in pregnancy; secretes HCG; presence indicates a positive test.

(b)   Progesterone prevents development of Graafian follicle; oestrogen inhibits FSH.

Ecology

1 (a)   All organisms/populations in a given habitat.

(b)   Diagram showing energy content at each trophic level, within a food chain.

(c)   Change in communities over time due to changing environmental conditions caused by community present.

2 (a)   Energy flows in one direction from primary producer upwards.

(b)   Energy lost at each trophic level; insufficient energy remains.

3 Inverted; represents standing crop; phytoplankton population fluctuates greatly.

4 (a)   Named greenhouse gas; combustion of fossil fuels; traps long-wave radiation/re-radiation; climate change; melting of polar ice cap; flooding; species migration.

(b)   Named gas; aerosols/refrigeration/supersonic aircraft; ozone broken down to oxygen; increased penetration of UV radiation; skin cancer/cataracts/crop damage.

(c)   Named acidic gases; combustion of fossil fuels; acidifies precipitation; decline in forests; acidification of freshwater; chemical weathering.

Exam Practice: Answers

1

Carbohydrate	Insoluble	Forms fibrils	Reducing sugar	Polymer of α-glucose
Starch	✔	✗	✗	✔
Cellulose	✔	✔	✗	✗
Maltose	✗	✗	✔	✗
Glycogen	✔	✗	✗	✔

2 (a) Columnar epithelial cell.

(b) *Two of*: many mitochondria – to produce ATP for active transport; microvilli – to increase surface area; many vesicles – showing endocytosis.

(c) An aggregation of similar cells that perform a particular function or functions.

3 external; upwards; outwards; diaphragm; flattened; increases; volume; decrease; air.

4

Vitamin	Solubility	Function	Deficiency
A	fat	vision	**night-blindness**
C	**water**	**connective tissue**	scurvy
D	fat	absorbs calcium	rickets
B_1	**water**	release of energy	**beri-beri**

5 (a) A = xylem; B = phloem; C = endodermis; D = cortex.

(b) Water moves into the root hair cells by osmosis / down a water potential gradient; moves across the cortex through the symplastic, apoplastic and vacuolar routes; until it reaches the endodermis where the Casparian strip prevents the apoplastic pathway and water moves into the symplastic pathway; water moves into the xylem tissue by osmosis.

(c) Hollow central lumen for movement of water and heavily lignified walls for support of water under pressure.

6 (a) (i) Red blood cells/erythrocytes.

(ii) Biconcave; no nucleus; increase surface area to pick-up/carry oxygen.

(b) (i) Hb can carry four O_2 molecules; Hb saturated at high partial pressure of O_2 in lungs; O_2 unloaded at lower partial pressure at tissues; later O_2 released more readily than first; haemoglobin most unstable when saturated.

(ii)   Increase.

(iii)   Increased dissociation of O_2; provides active tissue with more O_2 /enhances normal dissociation.

7 (a)   active site/description; forms steric relationship with substrate; other molecules unable to fit site.

   (b)   (i)   No permanent active site on enzyme; active site forms in the presence of enzyme.

         (ii)   Inhibitor same shape/charge as substrate; binds reversibly with active site.

   (c)   (i)   Volume of O_2 produced; in a given time.

         (ii)   Constant temperature; constant enzyme concentration/mass of potato; constant substrate concentration.

8 (a)   (i)   All organisms/populations in a given habitat.

         (ii)   Diagram showing mass of all organisms at each trophic level in a habitat/food chain.

         (iii)   Group of freely interbreeding individuals of the same species; living together in a given habitat.

   (b)   (i)   Mass/unit area; energy/unit area/year.

         (ii)   9.68%

         (iii)   Energy lost in respiration; as heat; not all organisms are eaten/digested; material passes to decomposers.

9 (a)   (i)   A – Stigma – receive/capture pollen; B – Anther – produce/release pollen; C – Ovule – forms the female gamete.

         (ii)   Insect pollination: pollen from anthers picked up by insect; transferred to stigma (of another plant).

   (b)   (i)   Pituitary gland; ovaries.

         (ii)   Chemical produced at one site/gland; having an effect at another site/target.

   (c)   (i)   Maintains endometrium; releases FSH; releases LH.

         (ii)   Development of the follicle; secretion of oestrogen.

10 nucleotides; deoxyribose; phosphate; nucleolus; messenger; nuclear envelope; transfer; amino acids; transcription; translation.